Beyond the Limit: Pushing Boundaries of Miniaturization with Atomic Precision

Jemison

TABLE OF CONTENTS

CHAPTER 1. Introduction

1.1 Motivation: Selectivity in Etching of Bulk Metal Films

Nanofabrication technologies provide unique advantages for applications including semiconductor processing, electrocatalysis, and energy storage.[1] With the downscaling of dimensions in various applications, especially semiconductor manufacturing, the need for nanoscale films has grown considerably. In recent years, researchers have been developing deposition and etching processes capable of providing atomic-level precision while maintaining film uniformity.[2,3] Atomic layer deposition (ALD) is of great interest because ALD involves assembly of metal monolayers one atomic layer at a time. ALD using electrochemical methods has been developed for copper (Cu)[4] and cobalt (Co).[5] Electrochemical ALD (e-ALD) provides unique advantages including atomically precise fabrication of metal films while utilizing benign liquid-phase chemistry.

The 'reverse' process of e-ALD, referred to as electrochemical atomic layer etching (e-ALE), has drawn attention from the semiconductor industry for its ability to remove material in a surface-limiting, layer-by-layer manner.[6,7] Modern microprocessors consist of billions of transistors connected through current-conducting Cu interconnects. To obtain faster microprocessors with more functionalities, more transistors need to be packaged in a single chip, and therefore the interconnect size must be reduced. Downscaling of interconnects below 10 nm has met challenges in the last decade. In the 'dual-damascene' process for fabricating nano-interconnect structures, high-aspect-ratio features called 'vias' and 'trenches' are first etched in a dielectric (SiO$_2$) layer, followed by vapor-phase deposition of a thin TaN barrier layer and then electrodeposition of Cu that fills up the vias and trenches. When a Cu-filled trench is in close proximity (<10 nm) to a via belonging to the adjacent layer as shown in Fig. 1.1, leakage current may flow between adjacent interconnects and cause electrical shorts. To increase the distance between adjacent interconnect structures during the fabrication of sub-10 nm scale nanoelectronics, atomic layer etching is an attractive technique for recessing the trench with atomic-level precision.[8]

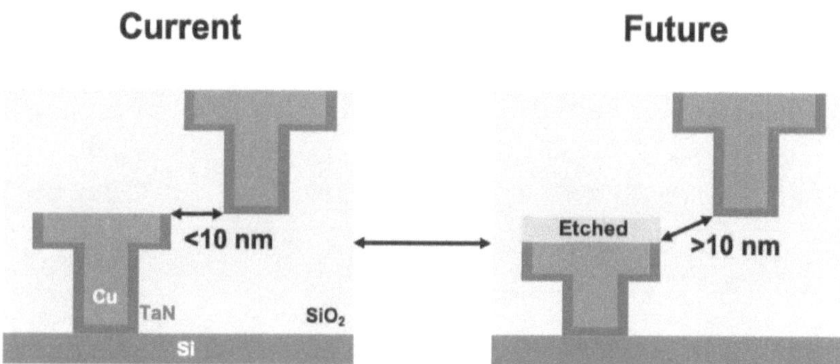

Figure 1.1. Schematic of interconnect fabrication process benefiting from atomic layer etching. The distance between adjacent interconnect structures can be increased from recessing the trench, thereby preventing leakage current and electrical shorts.

ALE of ruthenium (Ru) is also desired in nanoelectronics fabrication. There is growing interest in the industry to utilize a Ru barrier layer sandwiched between the deposited Cu and the underlying patterned silicon (Si) substrate. Ru is an attractive alternative to the conventional barrier layer materials – tantalum (Ta) and tantalum nitride (TaN), because Ru has low bulk resistivity (7.1 $\mu\Omega$-cm) compared to Ta (14 $\mu\Omega$-cm) and

TaN (~200 $\mu\Omega$-cm), high work function (4.7 eV), and low solubility in Cu.[9-11] The last step in the 'dual-damascene' process is to remove excess metals and planarize the surface via chemical mechanical polishing (CMP). During CMP, Ru is polished away at a slower rate than Cu because Ru is nobler than Cu, thus leaving behind Ru 'ears' that stick out of the surface as shown in Fig. 1.2. The Ru 'ear', when in close proximity (< 10 nm) to another interconnect structure, may cause leakage current and dielectric damage, thereby compromising device performance.[12-14] Thus, there is a critical need for Ru atomic layer etching, i.e., to removes Ru 'ears' and increase the spacing between adjacent interconnects.

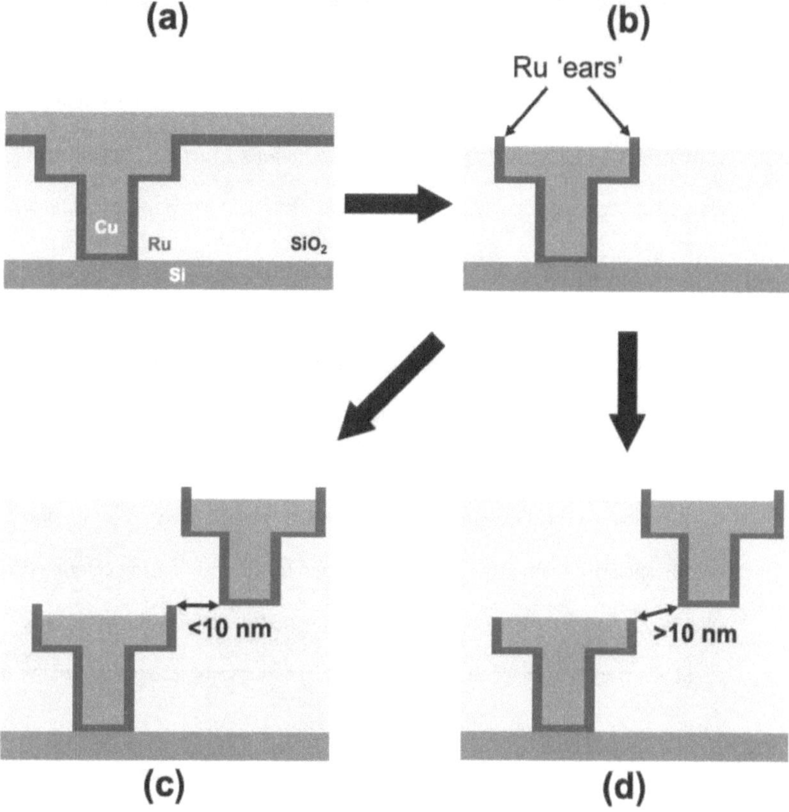

Figure 1.2. Schematic of interconnect fabrication process benefiting from atomic layer etching: (a) Filling of via and trench by Cu electrodeposition; (b) Ru 'ears' formed on the planar surface as a result of slower polishing (CMP) of Ru than Cu; (c) Sub-10 nm distance between Ru 'ear' and the adjacent interconnect may lead to leakage current and dielectric damage; (d) Increased distance between adjacent interconnects from etching of Ru 'ears' prevents leakage current.

1.2   Prior Work on Atomic Layer Etching

Atomic layer etching has been extensively studied. Table 1.1 summarizes several current atomic layer etching techniques for various materials. Vapor-phase, plasma-assisted etching is compared with liquid-phase etching in terms of their hallmark features (etch rate) and disadvantages.

Plasma-assisted atomic layer etching of silicon (Si) and silicon dioxide (SiO_2), as well as high dielectric constant oxides such as aluminum oxide (Al_2O_3), have been demonstrated. These processes typically consist of two surface-limiting steps: (*1*) surface modification and (*2*) removal of the reactive surface layer. For instance, surface modification of Si was achieved by chlorinating the Si surface using chlorine (Cl_2) gas or plasma-generated fluorocarbon ions to form a reactive surface layer. During this step, Si–Si bonds are broken for Cl to bond with Si.[15] The subsequent step removed this modified surface layer with argon ion bombardment. However, nonideal behavior could occur in both steps of the process. Chemical etching of Si during the chlorination step and sputtering of Si during the removal step are both disadvantages associated with plasma etching.

Table 1.1. Summary of current atomic layer etching techniques.

Method	Material	Etch Rate (Time)	Disadvantages	Source
Plasma-assisted Etching	Si	0.05–0.25 nm/cycle (0.6 min/cycle)	Undesired chemical and physical etching of Si; sub-atomic-layer etching; roughness amplification	Refs. 16–18
	SiO_2	60–760 nm/min	Incident ions for surface modification not completely removed during etching	Refs. 18–21
	Al_2O_3	0.01–0.17 nm/cycle (3.6 min/cycle)	Surface roughness amplification due to partial coverage of adsorbed reactive molecules	Ref. 22
	Cu	2.4–380 nm/cycle (~4 min/cycle)	Surface contamination by plasma-formed, nonvolatile etch byproduct; inconstant etch rate with increasing number of cycles	Refs. 23–29
	Ru	140–950 nm/min	Etching hindered by RuO_2 formed on Ru surface	Refs. 30–33
Wet Etching	Cu	1–5 nm/cycle (0.4–2.4 min/cycle)	Not surface-limiting	Ref. 8
	Ru	0.1–4.5 nm/min	Surface roughness amplification induced by etching	Ref. 34
	Co	1–3 nm/cycle (0.3–0.5 min/cycle)	Roughness amplification; simultaneous removal of TiN barrier	Ref. 35
	Ge	0.05–118 nm/min	Inconsistent etch rate; surface roughness amplification	Ref. 36

Plasma-assisted etching has also been applied to etching of metals. Conventional plasma-assisted processes for Cu etching require high temperatures (>200 °C) to volatilize the etch products.[6,23] Cu etching processes at room temperature have been developed in the last decade, with Cu subjected to two types of plasmas sequentially.[27,29] The first plasma (Cl_2 or HBr) aimed to oxidize the Cu surface, and the second plasma (H_2 or Ar) volatilized the oxidized surface layer. Although Cu etching with sub-nanometer resolution was achieved,[29] the etch rate still did not satisfy the requirements for atomic layer etching. Other disadvantages associated with plasma etching include the generation of nonvolatile or toxic byproducts and the need for high-temperature vacuum environments.[6,23,30,33,37]

Wet etching utilizing liquid-phase precursors has been investigated for recessing the Cu interconnect structure.[8] Cu was first oxidized by diluted hydrogen peroxide and then removed by ultra-diluted aqueous hydrogen fluoride. A stable recess etch rate of 1.0 nm/cycle corresponding to approximately four atomic layers of Cu was obtained within 15 cycles in 22.5-nm patterned Cu structure. The etch rate of Cu in the trench was higher than that of the blanket Cu due to a higher density of grain boundaries where the oxidation reaction preferentially occurred. Therefore, cycle time needs to be adjusted for tuning the Cu recess in a trench structure. Other current wet etching processes are also not surface-limiting, i.e., they lack selectivity in etching towards the metallic surface.

In all aforementioned studies, etching did not guarantee removal of bulk metal films in an atomic layer-by-layer manner. Such atomic-scale precision in etching is required to avoid surface roughness amplification. Therefore, an electrochemical approach for performing atomic layer etching is appealing due to its atomic-scale control over the etch rate using benign liquid-phase chemistry. It was once argued that a true atomic layer etching process could not be developed due to the lack of selectivity in etching of bulk metal films.[38] In this book, unprecedented true atomic layer etching is enabled through successful demonstration of surface oxidation of bulk metal films followed by selective etching of the oxidized monolayer.

1.3 Strategy: Surface Oxidation of Metals for Enabling Novel Electrochemical Atomic Layer Etching Processes

A benign electrochemical approach is utilized to achieve true atomic layer etching. In e-ALE processes, the surface monolayer of bulk metal films is oxidized before the oxidized monolayer is removed as shown in Fig. 1.3, thereby providing the selectivity critically desired in atomic layer etching. Compared to plasma etching, e-ALE is attractive for several reasons: (*i*) e-ALE utilizes liquid-phase chemistry and thus reduces the system complexity; (*ii*) e-ALE generates stable aquo-complexes which do not contaminate the metal surface; and (*iii*) e-ALE yields a uniform and constant etch rate of one monolayer per cycle with no significant surface roughness amplification.

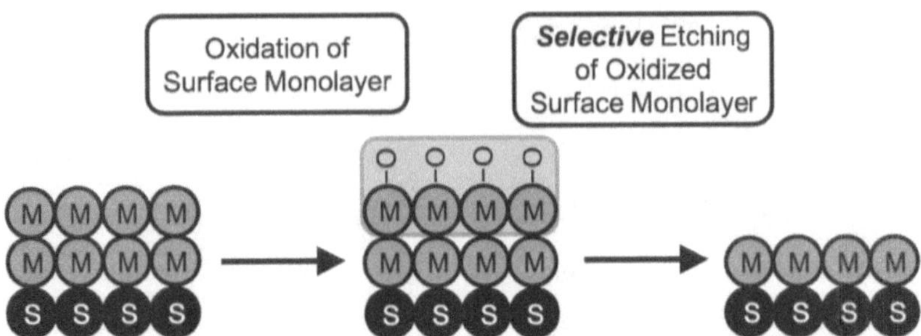

Figure 1.3. Schematic representation of electrochemical atomic layer etching processes for metals. S – substrate, M – metal, M–O – metal oxide.

The key to selectivity is to oxidize only the surface monolayer and not the underlying metal film. It is known that a stable chemisorbed monolayer may form on transition metal surfaces since oxygen species (atomic oxygen, sulfur, or hydroxyl groups) tend to adsorb on metal surfaces in aqueous solutions.[39,40] This intermediate surface monolayer precedes the formation of thicker surface oxides. For most metals, the oxidized surface monolayer is more resistive (10^{-5}–10 Ω-cm) than the metal (10^{-6}–10^{-5} Ω-cm), thus protecting the underlying metal from further oxidation.[11,41]

The metal cations in the oxidized surface monolayer are then subjected to complexation with the etchant species in the aqueous solution.[42–44] The change in Gibbs free energy (ΔG) for this complexation process is negative indicating spontaneous complexation. The complexes are stable and soluble in water, thus removing the surface metal atoms from the bulk metal film. The selectivity in etching of only the surface monolayer can be repeated to achieve layer-by-layer etching with atomic precision.

1.4 Objectives

Motivated by the limitations of currently available etching techniques and the unavailability of true atomic layer etching, the specific objectives of the present work are as follows:

(*i*) To develop a novel electrochemical atomic layer etching (e-ALE) process for removing Cu in a surface-limiting, layer-by-layer manner.

(*ii*) To perform thermodynamic studies of the surface-limited sulfidization of Cu – a key step in e-ALE of Cu, to ensure minimal surface roughness amplification during etching.

(*iii*) To develop a novel electrochemical ALE process for noble metals, e.g., Ru, enabled by surface oxidation of Ru.

Chapter 2 shows a novel 'two-cell' electrochemical ALE process for Cu. The etch rate and post-etch surface roughness are investigated using electroanalytical techniques such as cyclic voltammetry and anodic stripping coulometry in combination with surface characterization technique such as atomic force microscopy (AFM).

Chapter 3 presents thermodynamic studies for surface-limited sulfidization of Cu. Optimal Cu sulfidization potential is determined allowing the formation of a near-complete cuprous sulfide (Cu_2S) surface monolayer that is selectively removed during the subsequent etching step. At a Cu_2S surface coverage close to one, amplification of surface roughness can be suppressed as confirmed by underpotential deposition of Zn (Zn_{UPD}) on Cu.

Chapter 4 shows a 'one-cell' electrochemical ALE process for ruthenium developed by extending the approach in Chapter 3 to noble metals, e.g., ruthenium (Ru). Etching of Ru is confirmed via sheet resistance measurements. The etch rate and post-etch surface roughness are characterized using cross-sectional transmission electron microscopy.

Chapter 5 summarizes the key conclusions of this work and provides an outlook for future research.

CHAPTER 2. Electrochemical Atomic Layer Etching of Copper

2.1 Introduction

As described in Chapter 1, nanoscale copper (Cu) interconnect structures require recessing in the event that Cu interconnect comes into contact with adjacent interconnects causing leakage current and dielectric breakdown. An effective method to create the recess is through atomic layer etching (ALE). Conventional plasma-assisted Cu etching techniques utilizing vapor-phase precursors have met challenges in achieving layer-by-layer removal of materials without surface contamination by nonvolatile etch byproducts such as CuCl.[6,23,26,27] Although wet etching of Cu has been studied,[45–47] ALE using liquid-phase precursors and electrode potential manipulation has not been previously explored. Electrochemical ALE (e-ALE) is capable of providing atomic-scale control over etching in terms of etch rate and surface roughness while maintaining a contamination-free surface. In this chapter, a novel 'two-cell' e-ALE process for Cu is demonstrated. Surface-limited sulfidization of Cu and spontaneous etching of surface Cu are performed sequentially to etch multi-layers of Cu at a consistent etch rate while minimizing surface roughness amplification. The etch rate is determined using anodic stripping coulometry on Cu electrodes fabricated by electrochemical atomic layer deposition (e-ALD). The post-etch surface roughness is characterized via atomic force microscopy (AFM).

2.2 Experimental Details

2.2.1 Materials

Electrochemical ALE of Cu was studied using two electrochemical cells. The first cell used for surface-limited sulfidization of Cu contained 0.1 M potassium hydroxide (KOH, Fisher Chemical) and 0.5 mM sodium sulfide (Na_2S, Sigma-Aldrich), while the second cell used for spontaneous etching of surface Cu consisted of 2 M hydrochloric acid (HCl, Fisher Scientific). Electrolytes in both cells were prepared using analytically pure chemical reagents and de-aerated 18 MΩ-cm de-ionized (DI) water. Argon (Ar) gas was bubbled through the electrolytes for ~30 mins to remove dissolved oxygen. To study surface-limited sulfidization of Cu, a three-electrode cell was used consisting of a 100-nm Cu working electrode prepared by physical vapor deposition (PVD) of Cu onto a tantalum nitride (TaN)-coated Si wafer, a platinum (Pt) wire as the counter electrode, and a saturated Ag/AgCl reference electrode (Fisher Scientific).

2.2.2 Methods

Cyclic Voltammetry and Chronoamperometry Studies of Surface-limited Sulfidization of Cu.—Surface-limited sulfidization of Cu was studied using cyclic voltammetry (CV). Before electrochemical measurements, the Cu working electrode was rinsed first with ethanol and then with DI water before drying under a stream of nitrogen (N_2) gas. The cleaned Cu electrode was then immersed in 2 M sulfuric acid (H_2SO_4, Fisher

Scientific) for 1 min to remove surface Cu oxides. CV was performed in an electrolyte containing 0.1 M KOH and 0.5 mM Na_2S. A VersaSTAT 3 potentiostat was used for all electroanalytical measurements. All potentials reported below are with respect to the standard hydrogen electrode (SHE). CV measurements were conducted in the aforementioned three-electrode cell. A potential of -1.2 V vs. SHE was applied for 100 s to further reduce Cu oxides before the working electrode potential was scanned from -1.2 V to -0.6 V and back at a scan rate of 0.02 V/s. To achieve surface-limited sulfidization, chronoamperometry was performed on a Cu electrode at -0.75 V.

Electrochemical Atomic Layer Etching of Cu.—Cu e-ALE was performed on a Cu film fabricated using Cu e-ALD, which was previously developed by Venkatraman et al.[4] Cu e-ALD allowed e-ALE to be conducted on atomically flat Cu films. Cu e-ALD was performed for 10 cycles on a PVD-Ru electrode to achieve a ~2 nm Cu film with RMS surface roughness of ~0.2 nm. Surface-limited sulfidization of Cu was performed at -0.75 V vs. SHE in the above-mentioned sulfidization electrolyte. The sulfidized Cu electrode was then removed from the sulfidization electrolyte, rinsed with de-aerated DI water, dried under a stream of N_2 gas, and transferred to a de-aerated etching electrolyte containing 2 M HCl for 30 s. After the selective etching step, the Cu electrode was again rinsed with de-aerated DI water and dried under N_2 before it was transferred back to the sulfidization electrolyte for the subsequent e-ALE cycle. The surface-limited sulfidization and selective etching steps were repeated sequentially to achieve atomic-scale control over etching. The surface-limited signature of Cu sulfidization was confirmed by performing e-ALE on PVD-Cu electrodes.

Anodic Stripping Coulometry of Post-etch Cu.—Anodic stripping coulometry was performed to determine the amount of remaining Cu after various numbers of e-ALE cycles. The remaining Cu was electrochemically stripped in 50 mM H_2SO_4 by first scanning the electrode potential from open-circuit potential (OCP ≈ 0.4 V) to 0.6 V vs. SHE at a rate of 0.02 V/s and then holding the potential at 0.6 V until the stripping current decayed to zero. This stripping current measured was solely due to Cu dissolution since hydrogen evolution was thermodynamically prohibited at such potentials. The mass of remaining Cu after various number of e-ALE cycles was then calculated from the integrated stripping charge density measured via Faraday's law. A Dimension 3100 (Veeco Digital Instruments) atomic force microscopy (AFM) was used to measure film RMS roughness before and after e-ALE.

2.3 Results and Discussion

2.3.1 Cyclic Voltammetry and Chronoamperometry Studies of Surface-limited Sulfidization of Copper

The CV scan of a Cu electrode in an electrolyte containing 0.5 mM Na_2S and 0.1 M KOH (pH = 13.2) is shown in Fig. 2.1a (*blue*) in comparison with the background CV collected without Na_2S in the electrolyte (*red*). The IR drop in the electrolyte was estimated and found to be negligible (few mV). The positive limit of the scan potential was set at –0.6 V since surface Cu would be oxidized at more positive potentials.[48] The onset potential of hydrogen evolution reaction was near -1.0 V. An oxidation peak corresponding to bulk Cu_xS film formation was observed near -0.68 V in the anodic scan. A relatively small oxidation peak near -0.98 V corresponding to surface-limited sulfidization of Cu was also observed, consistent with previous studies that had reported sulfur adlayer formation on Cu at such potentials.[48,49] In the cathodic scan, a peak indicating reduction of bulk Cu_xS reduction was observed near -0.93 V.

The potential selected for surface-limited sulfidization of Cu is anodic with respect to the small oxidation peak near -0.98 V but cathodic with respect to bulk Cu_xS formation near -0.68 V. With the working electrode potential potentiostatically set at -0.75 V, the current transient obtained is shown in Fig. 2.1b (*left* axis). The initial unsteady state behavior can be attributed to transient diffusion and electro-nucleation processes.[50] After approximately 35 s, the current decays to nearly zero, indicating self-terminating behavior once the electrode surface is covered with a monolayer of Cu_xS. As seen in Fig. 2.1b (*right* axis), the integrated charge density for surface-limited sulfidization of Cu saturates at Q_{sls} ≈ 300 $\mu C/cm^2$, which is consistent with previous reports.[48] Assuming surface molar density of Cu(111) to be $N = 2.96$ nmol/cm^2 (calculated using the lattice constant of Cu from Ref. 51), the measured charge density Q_{sls} corresponds to a Cu oxidation state of $n = Q_{sls}/NF = 1.05$ where F is the Faraday's constant. This suggests the formation of a monolayer of Cu_2S ($x \approx 2$) on Cu.[48]

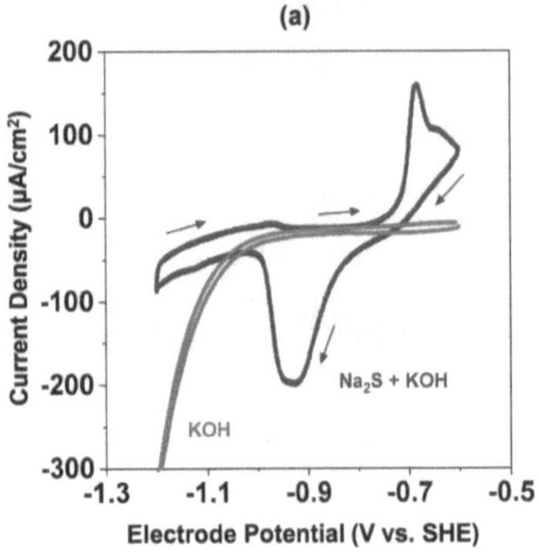

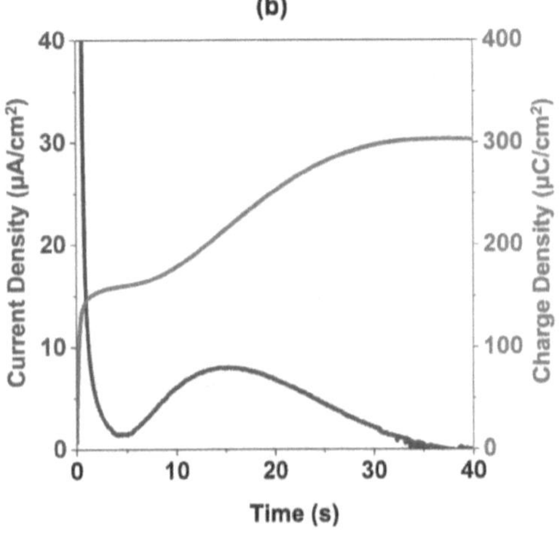

Figure 2.1. (a) Cyclic voltammogram of a PVD-Cu electrode in 0.5 mM Na$_2$S + 0.1 M KOH (*blue*) and in Na$_2$S-free solution of 0.1 M KOH (*red*). Scan rate was 0.02 V/s. (b) Current density (*left* axis) and integrated charge density (*right* axis) during sulfidization at –0.75 V vs. SHE.

35

2.3.2 Electrochemical Atomic Layer Etching of Copper Utilizing Surface-limited Sulfidization of Copper

A schematic representation of Cu e-ALE is depicted in Fig. 2.2. Cu e-ALE was achieved by sequentially repeating surface-limited sulfidization and selective etching of the formed Cu_2S monolayer by HCl:

(*1*) *Surface-limited sulfidization of Cu*: As described above, surface-limited sulfidization of Cu was performed in an alkaline electrolyte (pH = 13.2) containing 0.5 mM Na_2S. In this electrolyte, liquid-phase equilibrium favors the formation of hydrosulfide species (HS^-_{aq}).[49] At suitable electrode potentials, i.e., –0.75 V vs. SHE, Cu sulfidization reaction is facilitated:[48,52]

$$2Cu + [HS^-]_{aq} \rightleftharpoons [Cu_2S]_{solid} + [H^+]_{aq} + 2e^- \qquad [2.1]$$

When the surface Cu atoms are completely oxidized to Cu_2S, the reaction in Eq. 2.1 self-terminates as observed in Fig. 2.1b.

(2) *Selective etching of Cu$_2$S by HCl*: The Cu_2S layer formed during sulfidization was selectively etched by immersion in the HCl-containing etching electrolyte. The presence of chloride ions facilitates the spontaneous formation of stable aquo-chloro-complexes with Cu^{+1} from the Cu_2S monolayer:[43]

$$[Cu_2S]_{solid} + 6[Cl^-]_{aq} \rightleftharpoons 2[CuCl_3^{-2}]_{aq} + [S^{-2}]_{aq} \qquad [2.2]$$

The chloro-complex formed in Eq. 2.2 may also exist as $CuCl_2^-$ as described elsewhere.[53] In the absence of dissolved oxygen in the etching electrolyte, the underlying Cu^0 cannot be oxidized once the surface Cu_2S is etched via Eq. 2.2. When the surface Cu_2S is completely removed, Eq. 2.2 is also terminated.

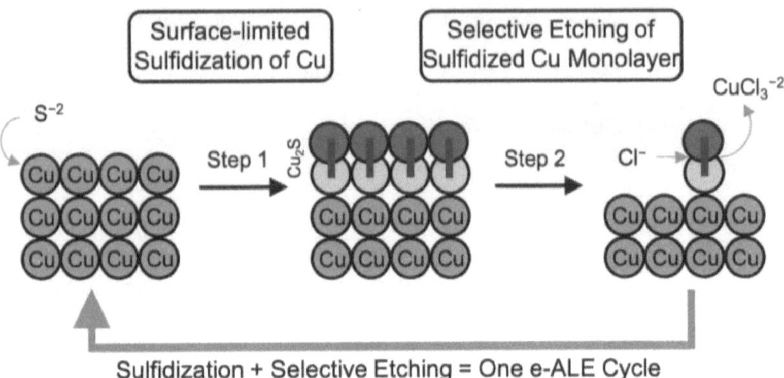

Figure 2.2. Schematic representation of the Cu electrochemical atomic layer etching process.

The surface-limited sulfidization signature was confirmed by performing e-ALE on PVD-Cu electrodes. As shown in Fig. 2.3, the oxidation current decayed to zero in approximately 35 s during the surface-limited sulfidization step of the 1ˢᵗ–5ᵗʰ cycles of e-ALE. During each e-ALE cycle, the integrated sulfidization charge density (Fig. 2.3 inset) remained fairly constant (except for the 1ˢᵗ cycle), indicating that the sulfidization and etching steps are quite repeatable.

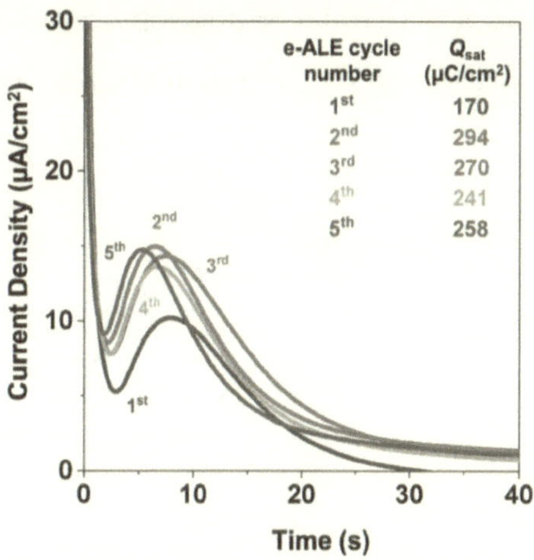

Figure 2.3. Chronoamperometry during the surface-limited sulfidization steps of the 1st–5th cycles of e-ALE.

2.3.3 Investigation of Etch Rate via Anodic Stripping Coulometry of Post-etch Copper Films

Anodic stripping coulometry was performed to quantify the amount of Cu before and after e-ALE cycles. As illustrated in Fig. 2.4, the integrated Cu stripping charge density (Q_0) corresponding to 10 atomic layers of Cu prepared by e-ALD was determined. The charge density (Q) for the remaining Cu after various e-ALE cycles was also determined. The normalized Cu stripping charge density (Q/Q_0), which represents the ratio of Cu remaining after e-ALE cycles, decreases linearly with increasing number of e-ALE cycles as shown in Fig. 2.5. The slope of the linear curve is equivalent to an etch rate of 0.103 (10.3%) per etch cycle. Since the original un-etched Cu film was prepared by depositing

38

10 atomic layers of Cu via e-ALD, removing approximately 10% of the film during each e-ALE cycle suggests an etch rate of approximately one Cu monolayer per e-ALE cycle.

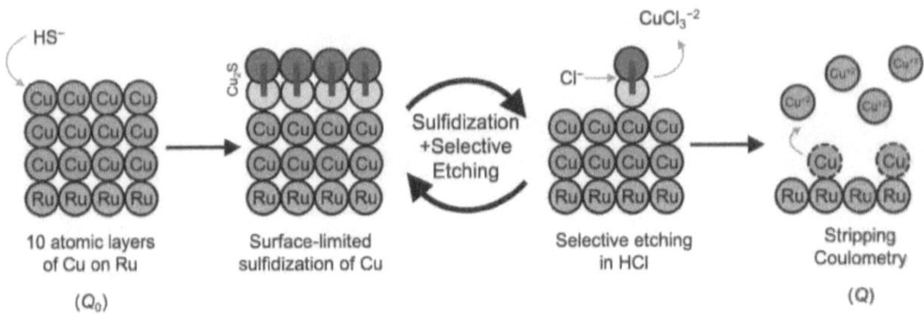

Figure 2.4. Schematic representation of anodic stripping coulometry on Cu before and after various Cu e-ALE cycles.

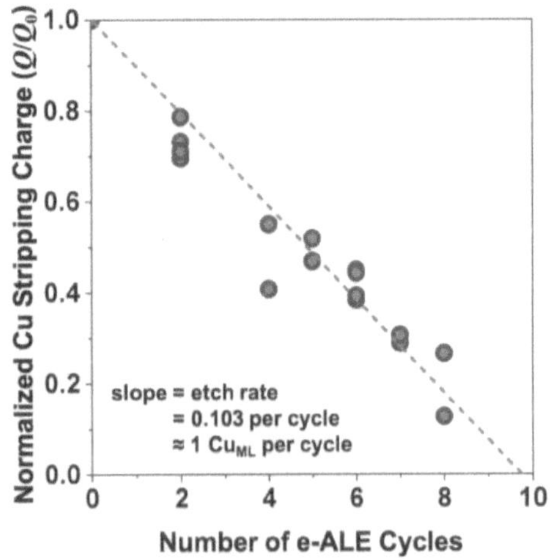

Figure 2.5. The normalized Cu stripping charge density (Q/Q_0) is a linear function of the number of e-ALE cycles. The slope of the linear curve corresponds to a Cu etch rate of approximately 10% (equivalent to approximately one Cu monolayer) per e-ALE cycle.

2.3.4 Characterization of Roughness Amplification during Etching via Atomic Force Microscopy

Surface roughness of Cu before and after e-ALE was characterized using atomic force microscopy (AFM). AFM line scans of the un-etched and etched Cu surfaces are shown in Fig. 2.6a. RMS roughness of the Cu surface is plotted in Fig. 2.6b as a function of number of e-ALE cycles performed. The RMS surface roughness after 4 and 6 e-ALE cycles remained relatively constant in the ~0.2 nm range compared to the un-etched Cu surface. In control experiments, Cu films subjected to Step (*2*) but not to Step (*1*) exhibited significant roughness amplification (0.92 nm).

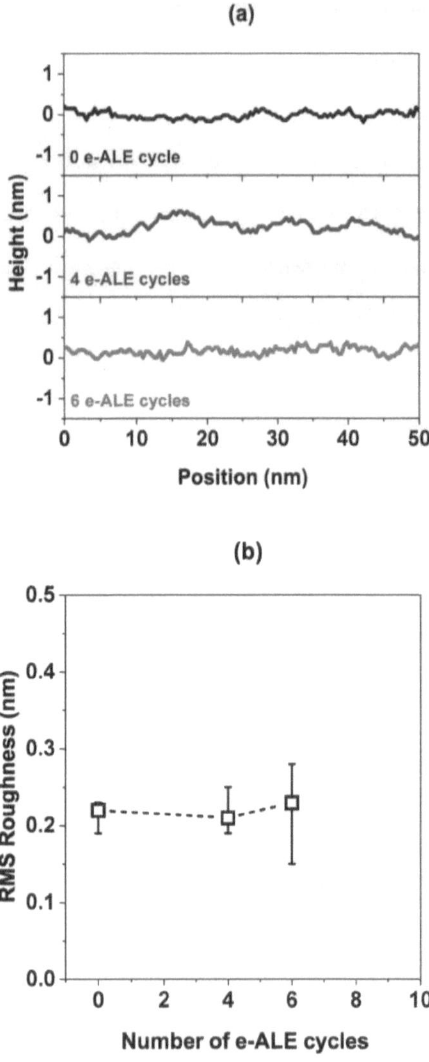

Figure 2.6. (a) AFM line scans of the original un-etched Cu surface and the etched Cu surfaces after 4 and 6 e-ALE cycles. (b) RMS roughness of the Cu surface as a function of number of e-ALE cycles performed. The error bars indicate maximum and minimum measured roughness values.

2.4 Conclusions

In this chapter, a novel electrochemical atomic layer etching process for Cu consisting of two steps is demonstrated: (*1*) surface-limited sulfidization of Cu to form a surface monolayer of Cu_2S and (*2*) selective etching of Cu_2S in HCl. Surface-limited sulfidization was studied using cyclic voltammetry and chronoamperometry where self-terminating behavior of the reaction was observed. An etch rate of approximately one Cu monolayer per e-ALE cycle was determined by performing various numbers of e-ALE cycles on pre-made Cu films with known amount of Cu. Additionally, smooth and uniform etching was confirmed by AFM. The e-ALE process presented in this chapter provides the selectivity, enabled by oxidation of the metal surface but not the underlying bulk metal, in removing only the surface monolayer during each cycle. The selectivity is crucial for the layer-by-layer removal of Cu with unprecedented atomic-scale control for potential applications in fabrication of advanced nanoelectronics.

CHAPTER 3. Thermodynamic Investigation of Surface-limited Sulfidization of Copper for Ensuring Minimal Surface Roughness Amplification during Etching

---

3.1 Introduction

In Chapter 2, the basic feasibility of a novel electrochemical atomic layer etching (e-ALE) process for copper (Cu) was demonstrated. The selectivity in layer-by-layer etching of Cu relies on oxidizing and then removing only the surface Cu monolayer. It is evident from the surface-limited sulfidization reaction (Eq. 2.1) that sulfidization must yield cuprous sulfide (Cu_2S) surface coverage (θ) close to unity to ensure the precise removal of a complete monolayer of Cu in one e-ALE cycle. If $\theta \ll 1$, sub-monolayer Cu_2S coverage would develop and this would result in non-uniform etching of Cu during e-ALE cycles. The non-uniformity would manifest in amplification of surface roughness during etching, which is typically undesired in semiconductor processing. Since the reaction in Eq. 2.1 involves electron transfer, the equilibrium Cu_2S coverage resulting from the sulfidization reaction is expected to be a function of the electrode potential. It is thus crucial to quantitatively characterize the relationship between θ and the applied electrode potential, and to develop guidelines for achieving optimal e-ALE whereby $\theta \approx 1$.

Thermodynamic investigations of surface-limiting electrochemical reactions are abundant in literature in the context of phenomena such as underpotential deposition (UPD) and electrosorption. For example, Kirowa-Eisner et al. used the Frumkin isotherm to model the equilibrium surface coverage obtained during UPD of lead on silver and gold substrates.[54] Bowles examined UPD of thallium on platinum where the Langmuir isotherm was found suitable for the extraction of parameters like the equilibrium constant and the surface saturation charge density.[55,56] Heiland et al. investigated electrosorption of benzene on platinum by applying the Frumkin isotherm and determining the equilibrium constant and the lateral interaction parameter associated with the adsorbed layer.[57] These prior studies provide a framework and fundamental basis for investigating the thermodynamic properties of surface-limited formation of a Cu_2S monolayer during Cu e-ALE. In this chapter, the equivalent adsorption isotherm for the surface-limited sulfidization reaction is investigated, and its implications to the optimal selection of electrode potentials during the e-ALE cycle are discussed. Amplification of surface roughness, as characterized by underpotential deposition of Zn (Zn_{UPD}) on Cu, can be suppressed when θ approaches unity, i.e., the optimal design and operation condition for Cu e-ALE.

3.2 Experimental Details

3.2.1 Materials

Cu e-ALE was performed on 100-nm thick polycrystalline Cu films fabricated by physical vapor deposition (PVD) on tantalum nitride (TaN)-coated Si wafers. The three-electrode electrochemical experimental setup has been described in Section 2.2.1. For surface-limited sulfidization of Cu (Eq. 2.1), the electrolyte was de-aerated 0.1 M potassium hydroxide (KOH, Fisher Chemical) with 0.5 mM sodium sulfide (Na_2S, Sigma-

Aldrich). For selective etching of Cu_2S (Eq. 2.2), the electrolyte was 2 M hydrochloric acid (HCl, Fisher Chemical).

Zn_{UPD} was performed in a three-electrode cell consisting of etched and un-etched Cu electrodes as the working electrode, a platinum (Pt) wire as the counter electrode, and a saturated Ag/AgCl reference electrode (Fisher Scientific). The electrolyte contained 0.1 M ammonium hydroxide (NH_4OH, Fisher Scientific) and 1 mM zinc sulfate heptahydrate ($ZnSO_4 \cdot 7H_2O$, Sigma Aldrich, >99.995% purity). Argon (Ar) gas was bubbled through the electrolyte for ~30 mins to remove dissolved oxygen.

3.2.2 Methods

Surface-limited sulfidization and electrochemical atomic layer etching of Cu.— The experimental procedure for Cu e-ALE has been described in detail in Section 2.2.2. Briefly, Cu e-ALE was performed in the two-step sequence shown by Eqs. 2.1–2.2 (Fig. 2.2). In Step (*1*), the applied sulfidization potential was either –0.74 V or –0.8 V vs. SHE, Additionally, surface-limited sulfidization experiments, i.e., only Step (*1*), were performed at potentials ranging from –0.94 V to –0.8 V vs. SHE. In these experiments, chronoamperometry was used to analyze the surface coverage of the Cu_2S adlayers formed.

Cu surface roughness characterization using Zn underpotential deposition.— Underpotential deposition of Zn (Zn_{UPD}) was performed to characterize the surface roughness of Cu electrodes before and after various numbers of e-ALE cycles. Cyclic voltammetry (CV) was employed to study Zn_{UPD} on Cu. The electrode potential was scanned from –0.3 V to –1 V at a scan rate of 0.02 V/s. The positive limit of the potential scan was set at –0.3 V to prevent copper oxide formation. The Zn_{UPD} monolayer was then electrochemically stripped by reversing the potential scan ending at –0.3 V vs. SHE. The stripping charge was used to quantify roughness because of the absence of parasitic reactions at –0.3 V. To underpotentially deposit Zn on etched and un-etched Cu electrodes, the electrode potential was held at –0.9 V for 60 s followed by anodic stripping at –0.3 V also for 60 s. Exposure of Cu substrates to air was minimized and all substrates were transferred from the etching electrolyte to the Zn_{UPD} electrolyte within 60 s.

3.3 Results and Discussion

3.3.1 Determination of Cuprous Sulfide Surface Coverage during Surface-limited Sulfidization of Copper

To determine the effect of sulfidization potential on Cu_2S surface coverage, sulfidization experiments were performed under potentiostatic conditions. The applied sulfidization potential was in the range from –0.94 V to –0.8 V vs. SHE. As shown in Fig. 3.1a, the sulfidization current decays down to nearly zero at a potential of –0.82 V indicating the surface-limiting behavior of Cu_2S formation. The charge density corresponding to sulfidization was calculated by integrating the current over time and was found to be 218 $\mu C/cm^2$. However, at potentials negative with respect to –0.82 V, i.e.,

–0.9 V, the current decayed to a cathodic steady-state current due to the presence of a background reaction as shown in Fig. 3.1b (*blue*). Without Na_2S in the electrolyte, i.e., in the absence of sulfidization, a cathodic steady-state current due to the hydrogen evolution reaction (HER) was detected at –0.9 V (*red*). Therefore, the true charge density associated with surface-limited sulfidization of Cu (Q_{sls}) at a given potential is that obtained from the sulfidization current transient (Fig. 3.1b, *blue line*) adjusted for the background HER charge density approximated using the cathodic steady-state current (Fig. 3.1b, *blue dashed line*). Knowing Q_{sls}, the resulting surface coverage of Cu_2S (θ) was determined:

$$\theta = \frac{Q_{sls}}{Q_{Cu_2S\text{-}ML}}\left(\frac{1}{r}\right) \qquad\qquad [3.1]$$

where $Q_{Cu_2S\text{-}ML} = 286\ \mu C/cm^2$ representing the theoretical charge density for a complete monolayer of Cu_2S.[51,58] The intrinsic Cu surface roughness factor (r) is defined and explained in the section below.

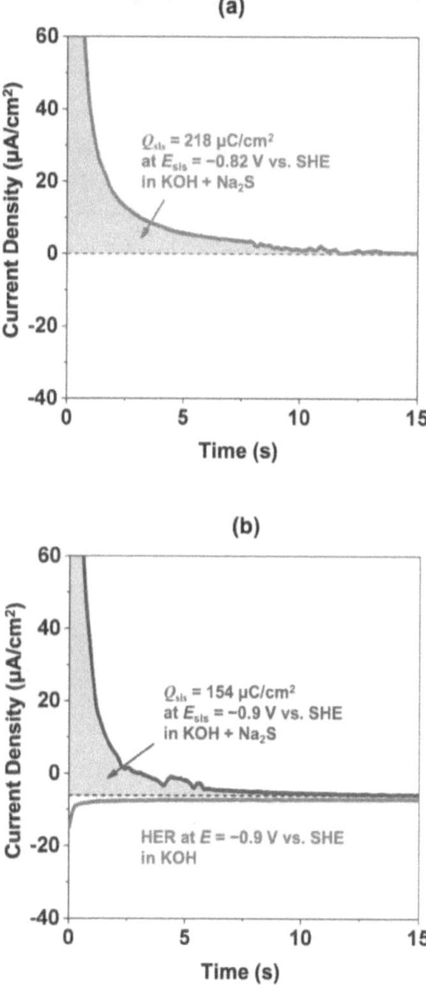

Figure 3.1. (a) Current transient (*green*) obtained on Cu at −0.82 V in 0.1 M KOH + 0.5 mM Na₂S. Q_{sls} represents the integrated charge density corresponding to surface-limited sulfidization. (b) Cathodic current transient (*red*) due to hydrogen evolution reaction obtained on a Cu electrode at −0.9 V vs. SHE in 0.1 M KOH. Sulfidization current transient (*blue line*) obtained on Cu at −0.9 V in 0.1 M KOH + 0.5 mM Na₂S. The cathodic steady-state current (*blue dash*) due to hydrogen evolution was used for background correction in determining Q_{sls}.

3.3.2 Surface Roughness Characterization of Etched and Un-etched Copper Films via Underpotential Deposition of Zinc

Cu surface roughness factor was determined via Zn_{UPD}. All starting Cu substrate materials used in e-ALE experiments have intrinsic roughness due to their polycrystalline nature and the PVD method used for their fabrication. Cyclic voltammogram on a typical sputter-deposited Cu film in an electrolyte containing 1 mM $ZnSO_4$ and 0.1 M NH_4OH is shown in Fig. 3.2a (*blue*). The IR drop in the electrolyte was estimated and found to be negligible (few mV). The background scan in the absence of $ZnSO_4$ is also shown (*red*). The broad reduction peak in the background scan at –0.65 V, corresponding to electrochemical hydrogen adsorption, was also observed in previous studies.[59,60] The cathodic peaks at –0.7 V and –0.85 V and the corresponding anodic peaks at –0.6 V and –0.45 V are due to Zn underpotential deposition and stripping, respectively. These peaks are absent in the background scan. To characterize the surface roughness of Cu before and after various numbers of e-ALE cycles, underpotential deposition of Zn was performed at –0.9 V vs. SHE followed by anodic stripping at –0.3 V. Electrolyte used was same as above: 1 mM $ZnSO_4$ and 0.1 M NH_4OH. During stripping, the Zn_{UPD} monolayer was completely removed as shown in Fig. 3.2b. Since hydrogen evolution was thermodynamically prohibited at –0.3 V, the stripping current measured was solely due to Zn_{UPD} oxidation. The integrated Zn_{UPD} stripping charge density (Q_{Zn}) is used to define the Cu surface roughness factor (r):

$$r = \frac{Q_{Zn}}{Q_{Zn\text{-ML}}} \qquad\qquad [3.2]$$

where $Q_{Zn\text{-ML}}$ ($= 340 \ \mu C/cm^2$) is the charge density associated with a Zn_{UPD} monolayer on atomically flat Cu.[61]

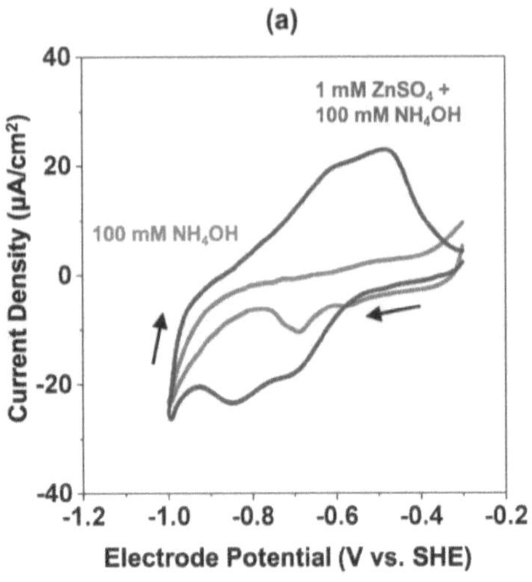

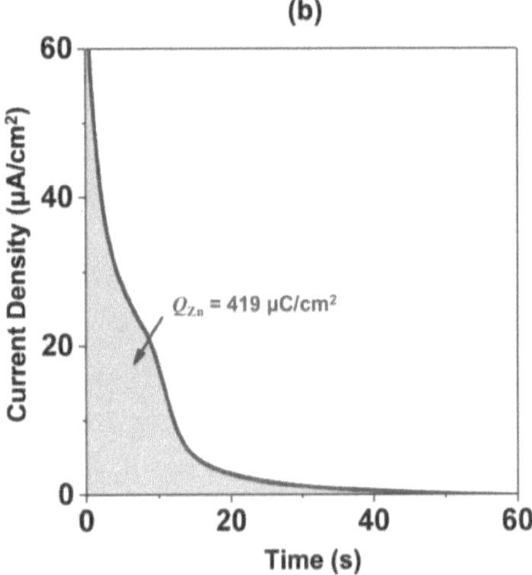

Figure 3.2. (a) Cyclic voltammogram of a Cu electrode in 1 mM ZnSO$_4$ + 0.1 M NH$_4$OH (*blue*) and in 0.1 M NH$_4$OH (*red*). Scan rate was 0.02 V/s. (b) Current density during potentiostatic (–0.3 V vs. SHE) anodic stripping of the Zn$_{UPD}$ monolayer formed at –0.9 V in 1 mM ZnSO$_4$ + 0.1 M NH$_4$OH electrolyte. The Zn$_{UPD}$ charge density indicates that the Cu electrode has a surface roughness factor of 1.23 per Eq. 3.2.

3.3.3 Thermodynamic Modeling of Surface-limited Sulfidization of Copper

During Step (1), the application of a sulfidization potential leads to the gradual development of Cu_2S surface coverage until equilibrium is reached wherein the surface Cu_2S equilibrates with the sulfide species in the aqueous electrolyte. At equilibrium, the net sulfidization reaction rate (current) reaches zero. The Cu_2S surface coverage (θ), measured by applying Eqs. 3.1–3.2 and aforementioned methods, was determined at various sulfidization potential (E_{sls}). The coverage was found to be a nearly linear function of E_{sls} in the intermediate coverage range ($0.3 < \theta < 0.7$) as can be seen in Fig. 3.3a. This linear behavior is characteristic of the Frumkin isotherm, which was used here to describe the dependence of Cu_2S surface coverage on the sulfidization potential (E_{sls}) and the concentration of HS^- (C_S):[62,63]

$$\frac{\theta}{(1-\theta)^2}\exp(f\theta) = K_0\exp\left[\frac{nF}{RT}(E_{sls} - E_{ref})\right]C_S \qquad [3.3]$$

where f is the lateral interaction parameter (positive values of f indicate repulsive interactions), K_0 is the equilibrium constant, n is the number of electrons transferred for Cu_2S formation ($n = 2$), F is the Faraday's constant, R is the ideal gas constant, T is the temperature, and E_{ref} is a reference potential. $(1 - \theta)$ representing the fractional coverage of exposed Cu surface is second order in Eq. 3.3, whereas θ is first order with respect to Cu_2S. Rearranging Eq. 3.3 gives

$$E_{\text{sls}} = \left(\frac{RT}{nF}\right)\left\{(f\theta) + \ln\left[\frac{\theta}{(1-\theta)^2 C_S K_0}\right]\right\} + E_{\text{ref}} \qquad\qquad [3.4]$$

In Eq. 3.4, E_{sls} vs. $\left\{(f\theta) + \ln\left[\frac{\theta}{(1-\theta)^2 C_S K_0}\right]\right\}$ exhibits a linear relationship with a slope equal

to $RT/nF = 0.013$ V. The Frumkin isotherm is plotted for intermediate surface coverages

with two fitting parameters, K_0 and f, as shown in Fig. 3.3b. For this plot, concentration of

HS$^-$ (C_S) was taken as the total sulfide concentration based on evidence that sulfide in

aqueous solutions is predominately hydrolyzed to hydrosulfide.[49] When $K_0 = 1$ M^{-1} and

$f = 24$, the expected slope (= 0.013 V) is obtained together with the Y-axis intercept of

$E_{\text{ref}} = -1.14$ V. Obtaining these thermodynamic parameters by fitting experimental data to

the Frumkin isotherm now allows us to determine a suitable potential at which a near-

complete Cu$_2$S monolayer can be formed and thus optimal e-ALE can be designed.

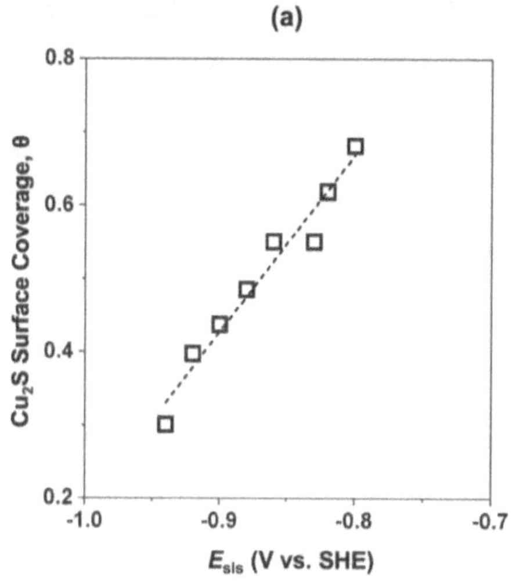

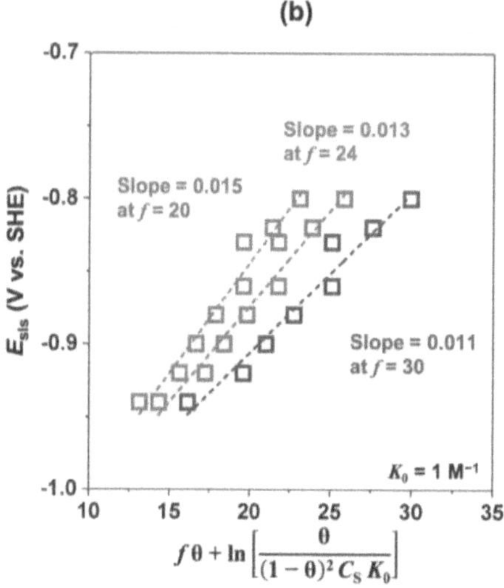

$$f\theta + \ln\left[\frac{\theta}{(1-\theta)^2\,C_S\,K_0}\right]$$

Figure 3.3. (a) Cu₂S surface coverage (θ) after Step (*1*) of e-ALE as a linear function of the surface-limited sulfidization potential (E_{sls}) at intermediate coverages ($0.3 < \theta < 0.7$). (b) The Frumkin isotherm Eq. 3.4 plotted at different f values and a fixed $K_0 = 1$ M⁻¹. At $f = 24$, the theoretically expected slope of 0.013 V (Eq. 3.4) is obtained.

54

Substituting the values of K_0, f and E_{ref} into Eq. 3.4 yields a function of θ with respect to E_{sls}. In this function (plotted in Fig. 3.4a), a sulfidization potential of -0.62 V is required theoretically to achieve a Cu_2S surface coverage of 0.99. However, when any sulfidization potential positive with respect to -0.74 V was applied, the reaction was found not to be surface-limiting as seen in Fig 3.4b. Even at -0.73 V, the current reaches a non-zero anodic steady-state value likely due to bulk sulfidization of Cu.[48,49] The increase in current after ~2 s in Fig. 3.4b is due to nucleation and growth phenomena on polycrystalline Cu electrodes as investigated by Moll et al.[50] Therefore, despite the theoretically desired E_{sls} of -0.62 V, practical limitations dictate that -0.74 V is a maximum allowable anodic sulfidization potential where the equilibrium Cu_2S surface coverage of 0.85 can be obtained (Fig. 3.4a). The surface roughness of the Cu electrode is now evaluated after e-ALE cycles where the surface-limited sulfidization is performed at -0.74 V.

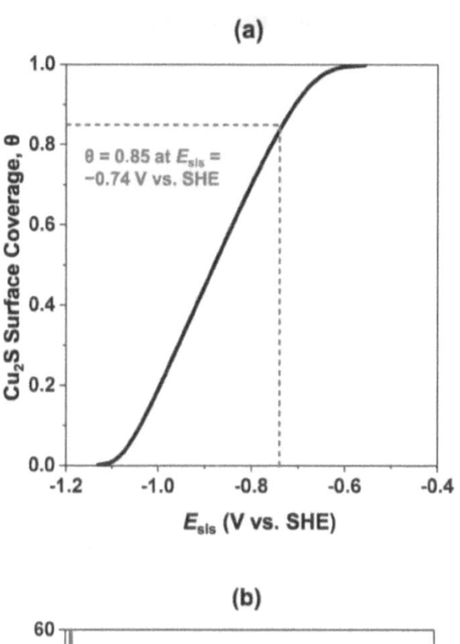

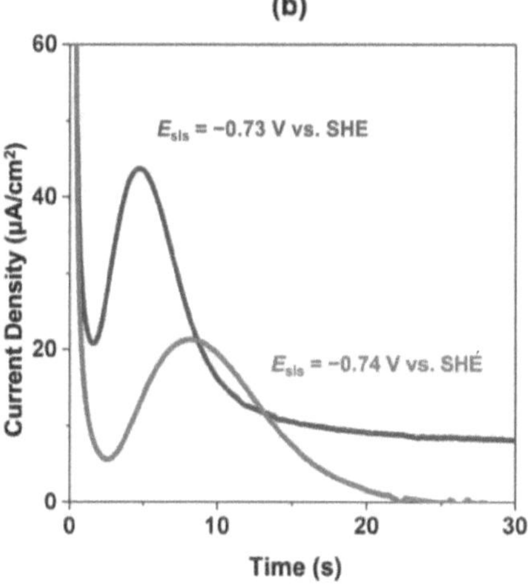

Figure 3.4. (a) Cu$_2$S surface coverage (θ) as a function of the surface-limited sulfidization potential (E_{sls}) following Eq. 3.4 with $K_0 = 1$ M^{-1}, $f = 24$, and $E_{ref} = -1.14$ V. (b) Sulfidization current transients obtained at -0.73 and -0.74 V vs. SHE in 0.1 M KOH + 0.5 mM Na$_2$S. Sulfidization at -0.74 V (*red*) is self-terminating but that at -0.73 V (*blue*) is not.

3.3.4   Investigation of Roughness Amplification after Etching

Zn_{UPD} on Cu followed by anodic stripping of the Zn_{UPD} adlayer was performed to characterize the surface roughness. The Zn_{UPD} stripping charge densities and the corresponding surface roughness factors (Eq. 3.2) for Cu subjected to various number of e-ALE cycles are shown in Fig. 3.5. The surface roughness factor is ~1.2 at $\theta = 0$ before any e-ALE cycles (*black*) were performed. During e-ALE of Cu, when the surface-limited sulfidization of Cu was conducted at –0.8 V resulting in $\theta = 0.68$ (*blue*), the surface roughness increased by 33% from 1.2 to 1.6 after 7 cycles of Cu e-ALE. However, when a near-to-complete monolayer of Cu_2S was formed at –0.74 V, i.e., $\theta = 0.85$ (*red*), the surface roughness of Cu did not evolve significantly and in fact was preserved at a value close to that of the initial un-etched substrate. This comparison demonstrates the importance of having a near-complete Cu_2S monolayer during e-ALE of Cu for achieving atomic layer-by-layer etching to preserve the surface smoothness.

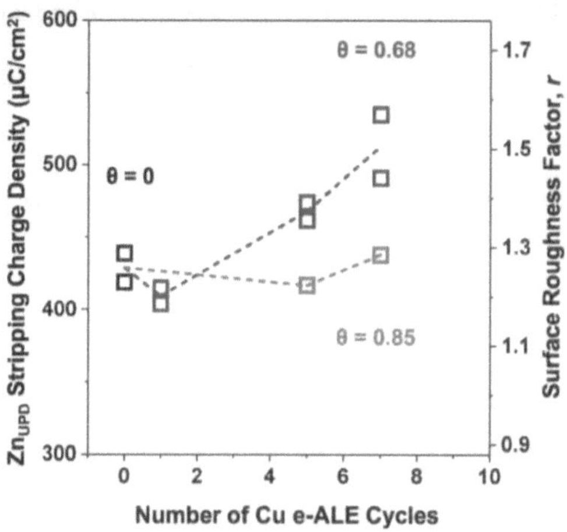

Figure 3.5. Zn_{UPD} stripping charge densities (*left* axis) and calculated Cu surface roughness factors (*right* axis) as a function of the number of e-ALE cycles performed. The surface roughness of post-etched Cu when $\theta = 0.68$ (*blue*) at $E_{sls} = -0.8$ V shows considerable amplification with cycles compared to Cu etched when $\theta = 0.85$ (*red*) at $E_{sls} = -0.74$ V. The surface roughness at $\theta = 0$ (un-etched Cu) is shown as reference (*black*).

3.4 Conclusions

Thermodynamic aspects of the surface-limited sulfidization of Cu to a Cu_2S adlayer during Cu electrochemical atomic layer etching were investigated in this chapter. Additionally, a thermodynamics-guided method for selecting the operating conditions, e.g., E_{sls}, during Cu e-ALE was provided. The equilibrium reached during surface-limited sulfidization was shown to follow the Frumkin isotherm, and the associated thermodynamic constants were extracted by comparison to experimental data. When a near-complete Cu_2S surface coverage ($\theta = 0.85$) is achieved at a sulfidization potential

$E_{sls} = -0.74$ V, the Cu surface roughness characterized by underpotential deposition of Zn is not significantly amplified during e-ALE. On the contrary, the Cu surface roughness increases significantly at $\theta = 0.68$ during sulfidization. This highlights the importance of careful control over the sulfidization potential during e-ALE for yielding a smooth post-etch Cu surface.

CHAPTER 4. Electrochemical Atomic Layer Etching of Ruthenium

4.1 Introduction

As reported in Chapter 1, there is growing interest in the industry on utilizing a ruthenium (Ru) barrier layer sandwiched between the deposited copper (Cu) and the underlying patterned silicon substrate. Non-uniform polishing of a dual-damascene structure can leave behind Ru 'ears' which potentially cause leakage current. Therefore, there is a critical need for processes that can etch Ru films essentially with atomic-scale precision. Although atomic layer etching (ALE) of Cu utilizing electrochemical methods has been recently achieved as discussed in Chapters 2–3, e-ALE of nobler metals, e.g., Ru, is yet to be demonstrated.

Previous efforts to etch Ru typically have involved vapor-phase, plasma-assisted techniques. These efforts have mainly focused on increasing the Ru etch rate and selectivity. Oxygen plasmas have been commonly used with the addition of chlorine gas to increase the etch rate.[30–32] Consequently, an etch rate greater than 140 nm/min and a Ru vs. SiO_2 etch selectivity of 6 could be achieved.[31,32] To further enhance the etch rate and selectivity, Ru was etched in ozone gas at a rate of 950 nm/min with a Ru vs. photoresist etch selectivity higher than 40. The large etch rates are useful for removing multilayers of Ru but undesired when it comes to atomically precise etching during nanoelectronics fabrication. Additionally, generation of toxic and reactive gaseous byproducts, e.g., RuO_4 can be problematic for practical applications.[30,33] Liquid-phase etching of Ru was also investigated in acidic solutions containing cerium (Ce^{+4}) species.[34] Although etch rates of sub-nm/min were achieved at Ce^{+4} concentrations smaller than 10 mM, drawbacks including surface roughness amplification and cerium-containing residues were observed by atomic force microscopy.

An alternative electrochemical atomic layer etching (e-ALE) process for Ru is presented in this chapter. Ru e-ALE utilizes benign liquid-phase electrochemistry in a single electrolyte to achieve layer-by-layer etching of Ru with atomic-level precision. Our three-step approach consists of: (*1*) surface oxidation of Ru, followed by (*2*) selective etching of the resulting ruthenium (III) hydroxide, Ru(OH)$_3$, monolayer through spontaneous complexation of Ru^{+3} with Cl$^-$, and lastly (*3*) removal of adsorbed Cl$^-$ on the Ru surface thus preparing it for subsequent ALE cycles. Electroanalytical studies including cyclic voltammetry and chronoamperometry are presented, which provide insights into the surface-limiting characteristics of Step (*1*) − Ru surface oxidation. Sheet resistance measurements on etched and un-etched Ru films and optical emission spectroscopy of the resulting electrolytes after etching both support Ru e-ALE. Ex-situ TEM facilitates determination of the Ru etch rate and Pb$_{UPD}$ confirms that Ru e-ALE minimizes surface roughness amplification during etching.

4.2 Experimental Details

4.2.1 Materials

Electrochemical ALE of Ru was studied in a three-electrode electrochemical cell with the Ru substrate serving as the working electrode, a platinum (Pt) wire as the counter electrode, and a saturated Ag/AgCl reference electrode (Fisher Scientific). Ru substrates were prepared by physical vapor deposition (PVD) of a ~5 nm Ru layer onto a TaN-coated silicon wafer. The electrolyte contained 0.5 M sulfuric acid (H_2SO_4, Fisher Scientific). Argon (Ar) gas was sparged through the electrolyte for ~30 mins to considerably lower the concentration of dissolved oxygen. A VersaSTAT 4 potentiostat (Princeton Applied Research) was used for all electroanalytical measurements. All potentials reported below are with respect to the standard hydrogen electrode (SHE).

4.2.2 Methods

Cyclic voltammetry studies of Ru surface oxidation.—Cyclic voltammetry (CV) was performed for characterizing the surface oxidation of Ru films. Before CV measurements, the surface of the Ru film was thoroughly cleaned in ethanol and in 18 MΩ-cm de-ionized (DI) water, followed by drying under a stream of nitrogen (N_2) gas. Ru surface oxides were cathodically reduced at an applied potential of –0.1 V vs. SHE for 30 s in a de-oxygenated 0.5 M H_2SO_4 electrolyte. The electrode potential was then scanned from 0 V vs. SHE to increasingly positive switching potentials (in the following order: 0.9, 1.0, 1.1, and 1.2 V vs. SHE), then returning back to 0 V at a scan rate of 0.02 V/s. The effect of the scan rate was investigated by performing CVs at 0.02, 0.04, 0.08, and 0.12 V/s.

Electrochemical atomic layer etching of Ru.—Ru e-ALE was performed in a three-electrode configuration (described above). The electrolyte contained 0.5 M H_2SO_4 and 50 μM sodium chloride (NaCl, Fisher Scientific). The e-ALE process for Ru consists of three steps: (*1*) surface oxidation of Ru at an applied potential of 0.95 V vs. SHE for 120 s, (*2*) spontaneous Ru^{+3}–Cl^- water-soluble complex formation under open-circuit conditions for 120 s, and (*3*) a potential hold at –0.1 V vs. SHE for 30 s to remove adsorbed Cl^- on the surface of the Ru film thereby preparing the substrate for the subsequent steps (*1*) and (*2*). Steps (*1*) to (*3*) were repeated cyclically to etch multi-layered Ru film in a true ALE mode. Etching was confirmed by comparing the sheet resistance of etched and un-etched Ru films. The sheet resistance was measured using a four-point probe (Alessi). The concentration of Ru in the e-ALE electrolyte after various numbers of e-ALE cycles was measured with inductively coupled plasma optical emission spectroscopy (ICP-OES, Agilent Technologies 700 Series). The Ru etch rate was determined by measuring the thickness of un-etched and etched Ru films using cross-sectional transmission electron microscopy (TEM).

Surface roughness characterization using underpotential deposition of Pb on Ru.— Underpotential deposition of Pb (Pb_{UPD}) was performed on un-etched and etched Ru films for surface roughness characterization. The electrolyte for Pb_{UPD} contained 0.2 mM lead perchlorate [$Pb(ClO_4)_2$, Acros Organics] and 10 mM perchloric acid ($HClO_4$, Fisher

Scientific). The electrolyte was sparged with Ar gas for ~30 mins to remove dissolved oxygen. The Ru electrode was transferred from the e-ALE electrolyte to the Pb$_{UPD}$ electrolyte within 60 s. To underpotentially deposit Pb on Ru, the electrode potential was scanned from the open-circuit potential (OCP) to –0.1 V vs. SHE at a scan rate of 0.02 V/s. The Pb$_{UPD}$ monolayer was then electrochemically stripped by reversing the potential scan to 0.45 V vs. SHE followed by a potential hold for 60 s. The Pb$_{UPD}$ charge density provided a measure of the electrochemically active surface area of the Ru film with increasing number of e-ALE cycles. Surface roughness comparison between un-etched and etched Ru films was also available via cross-sectional TEM.

4.3 Results and Discussion

4.3.1 Cyclic Voltammetry Studies of Surface Oxidation of Ruthenium

Cyclic voltammograms performed on a PVD-Ru electrode in a de-aerated electrolyte containing 0.5 M H_2SO_4 are shown in Fig. 4.1. The electrode potential was scanned from 0 V vs. SHE to increasingly positive switching potentials (in the order: 0.9, 1.0, 1.1, and 1.2 V vs. SHE) and back at a scan rate of 0.02 V/s. In the anodic scan direction, an oxidation peak was observed at 0.6 V vs. SHE corresponding to the formation of a $Ru(OH)_x$ monolayer via surface oxidation of Ru. This observation is consistent with previous studies on surface electrochemistry of Ru.[64–71] The current increased drastically at potentials anodic to 0.8 V vs. SHE corresponding to bulk oxidation of Ru.[68–71] In the cathodic scan direction, a peak appeared at around 0.4 V vs. SHE indicating reduction of

the oxidized surface Ru, whereas the peaks between 0.3 and 0.1 V corresponds to reduction

of the bulk Ru oxides. The significant increase in current at potentials cathodic to 0.05 V

vs. SHE was due to hydrogen evolution.

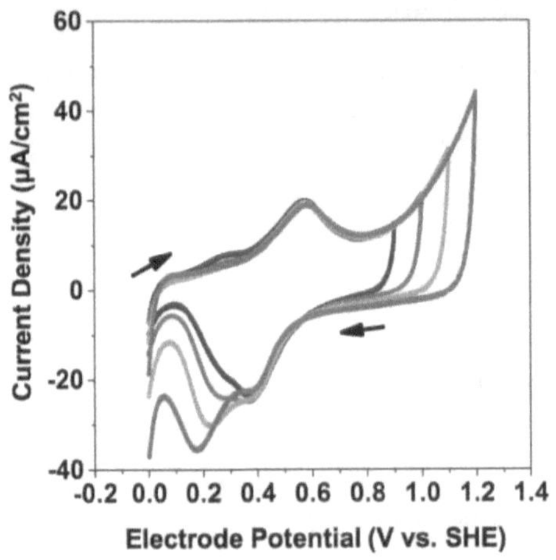

Figure 4.1. Cyclic voltammograms of Ru in 0.5 M H_2SO_4 starting from 0 V vs. SHE to increasingly positive switching potentials (in the order: 0.9, 1.0, 1.1, and 1.2 V vs. SHE) and back to 0 V. Scan rate was 0.02 V/s.

To demonstrate self-terminating characteristics of Ru surface oxidation, CV was

performed on a Ru electrode from 0 to 0.9 V vs. SHE at various scan rates (0.02, 0.04, 0.08,

and 0.12 V/s) as shown in Fig. 4.2a. The oxidation peak current density was found to be a

linear function of scan rate as shown in Fig. 4.2b. This linear relationship indicates that

surface oxidation of Ru is a self-terminating reaction, i.e., the reaction rate is suppressed

as the surface of Ru is gradually oxidized. For self-terminating electrochemical processes, the peak current density is known to depend on the scan rate,[54,72] which is observed in Fig. 5.2b. Oxidation of Ru is terminated once the Ru surface monolayer was oxidized. The surface-limiting characteristics of Ru oxidation are further investigated using chronoamperometry, as discussed in the next section.

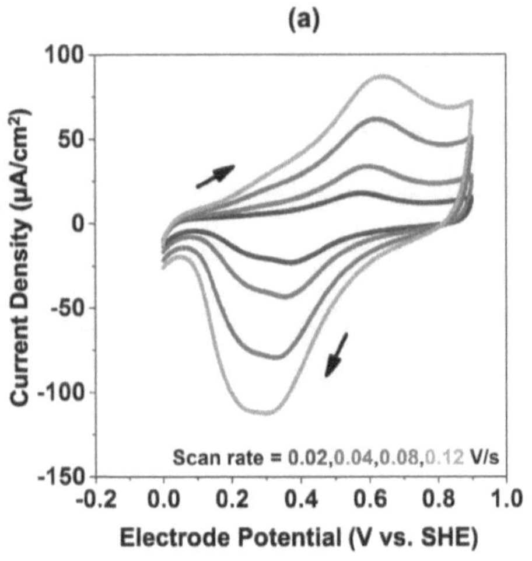

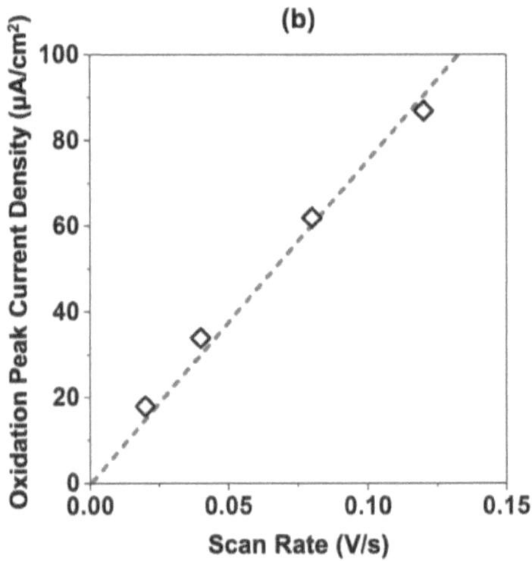

Figure 4.2. (a) Cyclic voltammograms of Ru in 0.5 M H_2SO_4. Scan rate was varied from 0.02 to 0.12 V/s. (b) Oxidation peak current density is a linear function of scan rate, indicating that surface oxidation of Ru is a self-terminating reaction.

4.3.2   Electrochemical Atomic Layer Etching of Ru Utilizing Surface Oxidation of Ruthenium

A schematic representation of Ru e-ALE is depicted in Fig. 4.3. The e-ALE electrolyte consisted of 0.5 M H_2SO_4 and 50 µM NaCl and was de-aerated via Ar purging. The Ru e-ALE process consists of three steps:

(*1*) *Surface oxidation of Ru*: The Ru surface was oxidized at an electrode potential of 0.95 V vs. SHE. The applied potential of 0.95 V is anodic to the Ru surface oxidation peak at 0.6 V but cathodic to bulk oxidation of Ru (Fig. 4.1 and 4.2). The transient response of the current (chronoamperometry) during the Ru surface oxidation step in the 1st and 2nd e-ALE cycles can be seen in Fig. 4.4. The current decayed to nearly zero after ~120 s indicating oxidation of Ru in a self-terminating manner. The integrated charge density saturated at ~700 µC/cm^2. Assuming a Ru(0001) surface which has a lower surface energy than other orientations,[73] its surface molar density (N) is ~2.62 nmol/cm^2. The measured saturation charge density (Q_{sat}) corresponds to a Ru oxidation state of $n = Q_{sat}/NF = 2.77$ suggesting the formation of a monolayer of Ru(OH)$_3$ ($x \approx 3$) on the Ru surface. At suitable potentials, i.e., 0.95 V vs. SHE, the formation of Ru(OH)$_3$ monolayer proceeded via the reaction:[74]

$$[Ru]_{ML} + 3H_2O \rightleftharpoons [Ru(OH)_3]_{ML} + 3[H^+]_{aq} + 3e^- \qquad [4.1]$$

(*2*) *Spontaneous Ru^{+3}–Cl^- complexation*: Following Step (*1*), the formed Ru(OH)₃ monolayer was selectively etched as Cl^- in the electrolyte facilitated spontaneous formation of stable aquo-chloro-complexes with Ru^{+3} in the Ru(OH)₃ monolayer. Thus, under open-circuit conditions, the following reaction is thought to proceed:

$$[Ru(OH)_3]_{ML} + [Cl^-]_{aq} \rightleftharpoons [RuCl^{+2}]_{aq} + 3[OH^-]_{aq} \qquad [4.2]$$

Although $RuCl^{+2}$ complex was the species featured in the reaction in Eq. 4.2, $RuCl_2^+$, $RuCl_3$, and $RuCl_4^-$ may also exist as previously reported.[42,75–81] The electrode potential transients under open-circuit conditions for the 1st and 2nd e-ALE cycles can be seen in Fig. 4.4. The electrode potential drifted in the negative direction reaching approximately 0.7 V vs. SHE after 120 s (close to the open circuit potential of the unetched Ru in the same electrolyte, *black* curve in Fig. 4.4) suggesting that the Ru(OH)₃ monolayer was selectively etched and the underlying Ru was exposed. Note that in de-oxygenated, Cl^- containing acidic electrolytes, Ru is stable and does not corrode.

(*3*) *Removal of adsorbed Cl^- on Ru*: The Cl^- ions in the electrolyte tend to adsorb on Ru at electrode potentials between 0 and 0.5 V vs. SHE.[82] The adsorbed Cl^- could inhibit surface oxidation of Ru in the subsequent e-ALE cycle. To eliminate the effect of Cl^- adsorption, the electrode potential was held at –0.1 V vs. SHE for 30 s. This potential is cathodic with respect to the potential range mentioned above where Cl^- adsorbs on Ru. Also, the applied potential of –0.1 V favors hydrogen evolution on Ru but not in copious amounts which would absorb into or block the Ru surface. The adsorbed Cl⁻ was removed from Ru surface allowing the surface oxidation step in the subsequent e-ALE cycle, as seen in Fig. 4.4. Step (*3*) was found to be crucial in achieving reproducibility of e-ALE as Step (*1*) in the 2ⁿᵈ e-ALE cycle would not yield adequate oxidation charge without Step (*3*) in the 1ˢᵗ cycle.

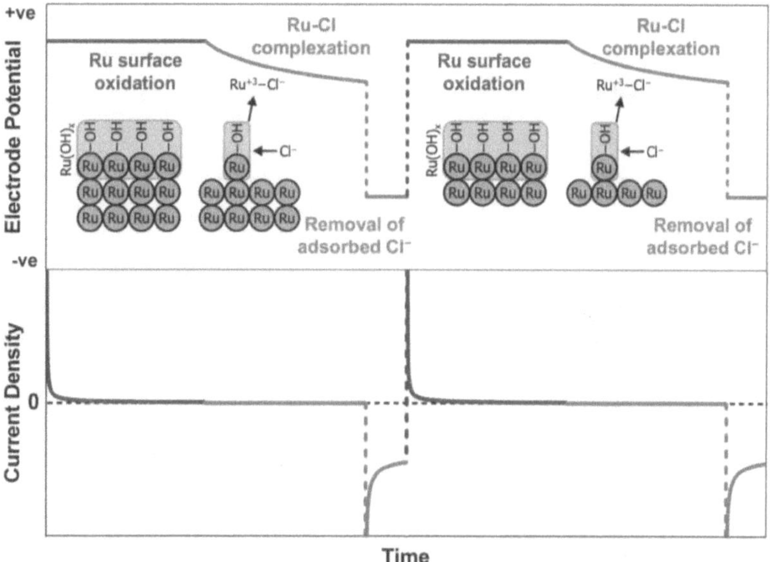

Figure 4.3. Schematic representation of the Ru e-ALE process comprising three steps: (*1*) potentiostatic surface oxidation of Ru at suitable electrode potentials; (*2*) selective etching of the formed Ru(OH)ₓ monolayer by Cl⁻ under open-circuit conditions; and (*3*) desorption of the adsorbed Cl⁻ on Ru at cathodic potentials so that subsequent e-ALE cycle can commence.

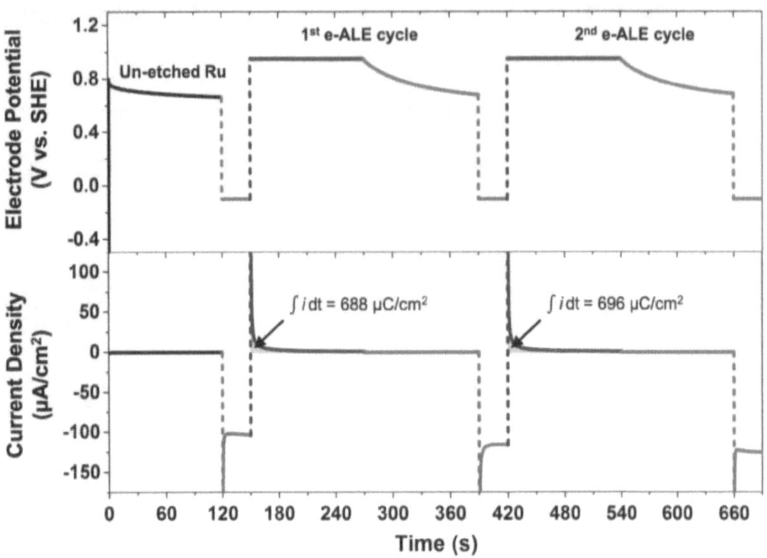

Figure 4.4. Transient response (experimental data) of the electrode potential and current density during the various steps involved in e-ALE of Ru (refer to schematic in Fig. 4.3). Electrolyte contained 0.5 M H_2SO_4 and 50 µM NaCl.

4.3.3 Investigation of Etch Rate via Cross-sectional Transmission Electron Microscopy

Steps (*1*) to (*3*) were repeated cyclically to achieve layer-by-layer etching of Ru. To confirm etching, the sheet resistance of etched and un-etched Ru films was measured using a four-point probe. Sheet resistance (R_s) is a function of film resistivity (ρ) and film thickness (t):

$$R_s = \frac{\rho(t)}{t} \qquad\qquad [4.3]$$

In the thin-film regime, the resistivity of Ru increases non-linearly with decreasing film thickness. If e-ALE provides a uniform and constant etch rate, i.e., a constant decrease in x per cycle, $\rho(t)$ would increase with more e-ALE cycles performed due to thin-film effects (electron scattering at interfaces and grain boundaries) on resistivity. Thus, in Eq. 4.3, sheet resistance R_s is expected to increase non-linearly with increasing number of e-ALE cycles as confirmed in Fig. 4.5. This increase in R_s confirms Ru etching. Etching of Ru is further verified by performing ICP-OES on the electrolytes before and after e-ALE. ICP-OES confirmed that the concentration of Ru in the e-ALE electrolyte was 13.5 ppb after 10 cycles of e-ALE, and it nearly doubled to 25.3 ppb after 20 cycles of e-ALE as expected.

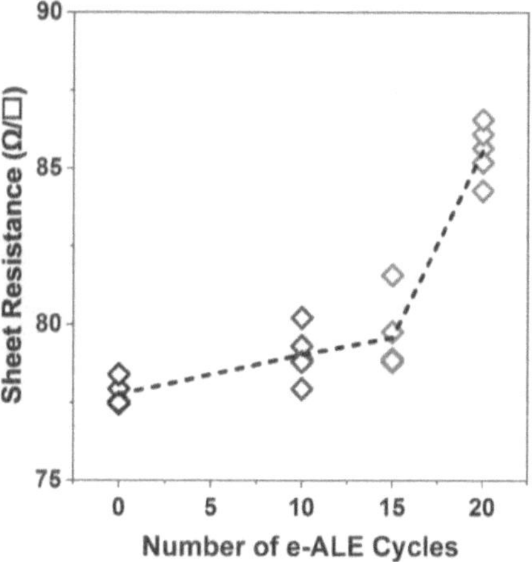

Figure 4.5. Sheet resistance of etched and un-etched Ru films as a function of the number of e-ALE cycles. A non-linear increase in the sheet resistance with increasing etch cycles is observed. This is due to the non-linear dependence of resistivity ρ on thickness at sub-10 nm dimensions.

The etch rate for the e-ALE process was determined by performing cross-sectional TEM (Fig. 4.6). In the TEM images, the Ru film thickness was measured at various locations, and plotted the thickness against the number of e-ALE cycles performed. As seen in Fig. 4.6, the thickness exhibits a linear relationship with the number of e-ALE cycles. The slope of this curve yielded an etch rate of 0.12 nm per e-ALE cycle. The Ru etched in one cycle is smaller in thickness than the inter atomic-layer spacing (0.214 nm) for Ru(0001).[83] This may be due to a few different reasons: (*i*) The Ru surface was not completely oxidized during Step (*1*) because the integrated charge density of 688–696 μC/cm^2 is somewhat lower than the theoretical saturation charge density for the three electron transfer ($n = 3$) surface oxidation of Ru ($nNF = 758$ μC/cm^2); and (*ii*) The oxidation state of Ru in the formed Ru(OH)$_x$ monolayer may be higher (e.g., $n = 4$) in accord with the thermodynamic analyses of Povar and Spinu.[44,80] Both factors could lead to an etch rate less than one atomic layer of Ru for every e-ALE cycle.

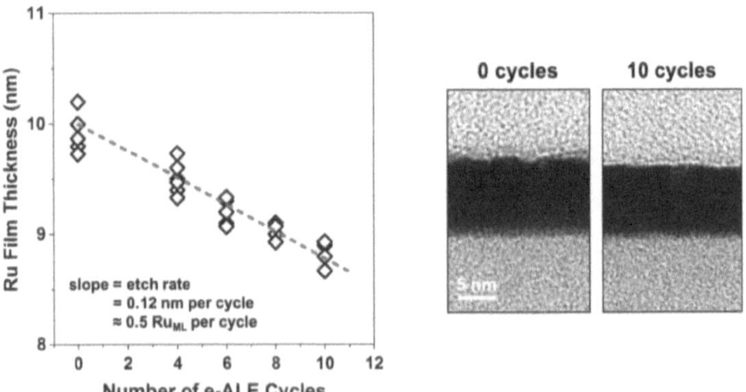

Figure 4.6. The Ru film thickness decreases linearly with the number of e-ALE cycles performed with a slope corresponding to a Ru etch rate of 0.12 nm per e-ALE cycle. Cross-sectional TEM images of the Ru film before and after 10 cycles of e-ALE again confirm etching.

In comparison to the previously developed Cu e-ALE process where the surface-limiting oxidation (sulfidization) step and the selective etching step are conducted in two separate electrolytes,[58] the capability of performing all three steps of the Ru e-ALE process in one electrochemical cell provides the following benefits: (*i*) electrode transfer between electrolytes in separate cells is not required, thus eliminating the possibility of oxidizing the electrode surface and improving overall process reproducibility; and (*ii*) all three steps can be performed in a continuous sequence without interruptions, which simplifies process scale-up.

4.3.4 Surface Roughness Characterization of Ruthenium via Underpotential Deposition of Lead

Characterizing the surface roughness amplification during Ru e-ALE to confirm true layer-by-layer etching of Ru is important as chemical and electrochemical etching methods tend to produce non-uniform roughened surfaces. After e-ALE of Ru, the Ru electrode was transferred from the e-ALE electrolyte to a de-aerated Pb$_{UPD}$ electrolyte containing 0.2 mM $Pb(ClO_4)_2$ and 10 mM $HClO_4$. The Ru electrode potential was scanned from OCP to -0.1 V vs. SHE at a scan rate of 0.02 V/s to facilitate Pb$_{UPD}$ on the surface of the Ru film as shown in Fig. 4.7. The underpotentially deposited Pb was subsequently stripped from the Ru surface in a linear scan from -0.1 to 0.45 V vs. SHE. For reference, background linear scan voltammograms collected in the absence of Pb^{+2} are also shown. The broad reduction peak in the background scan at $+0.25$ V vs. SHE, corresponding to reduction of Ru oxides formed during electrode transfer between electrolytes, was also observed in previous studies.[68,84,85] At more negative potentials (i.e., −0.1 V), the current surge was due to onset of hydrogen evolution. In the presence of Pb^{+2} in the electrolyte, Pb exhibited UPD on Ru at 0.06 V vs. SHE consistent with the prior work of Yu and Akolkar.[85] The charge density associated with Pb$_{UPD}$ was calculated by subtracting the background charge from the total charge (shaded region in Fig. 4.7). The Pb$_{UPD}$ charge density provides a measurement of the electrochemically active surface area of the Ru film as a function of the number of e-ALE cycles to which it is subjected. In Fig. 4.7b, the linear scan voltammograms collected after 0, 2, and 6 Ru e-ALE cycles practically overlapped with one another indicating minimal surface roughness evolution. The Pb$_{UPD}$ charge density remained relatively constant at 316–351 $\mu C/cm^2$ up to 6 e-ALE cycles as shown in Table 4.1. The charge density corresponds to the formation of a Pb monolayer on Ru.[72] Surface roughness comparison between un-etched and etched Ru films was also available via the cross-sectional TEM images (Fig. 4.6). These images, together with the Pb$_{UPD}$ studies, confirmed that surface roughness of the Ru film did not significantly change during the first 10 e-ALE cycles.

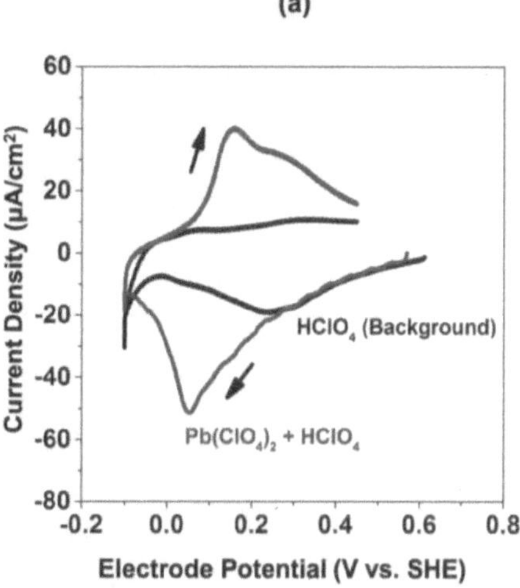

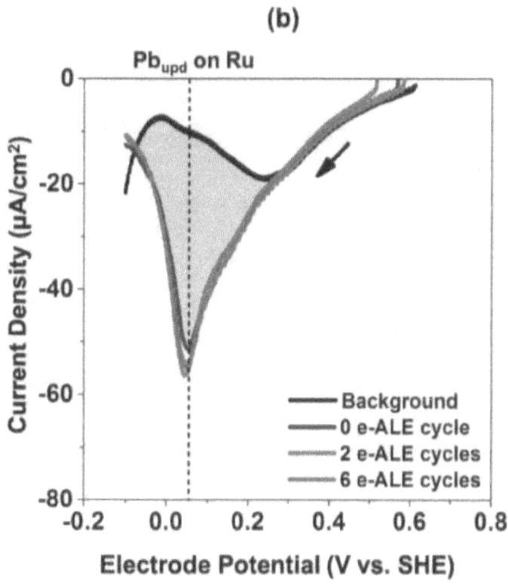

Figure 4.7. (a) Linear scan voltammograms of Ru in 0.2 mM Pb(ClO₄)₂ + 10 mM HClO₄ (*blue*) showing Pb_UPD on Ru in the cathodic scan direction and stripping of the underpotentially deposited Pb in the anodic scan direction. Background scan in 10 mM HClO₄ only (*black*) is shown for comparison. Scan rate was 0.02 V/s. (b) Linear scan voltammograms showing Pb_UPD on Ru after 0, 2, and 6 e-ALE cycles. The Pb_UPD charge density (shaded region) remains relatively constant up to 6 e-ALE cycles.

Table 4.1. Pb$_{UPD}$ charge density before and after e–ALE. The relatively constant Pb$_{UPD}$ charge density confirms minimal surface roughness amplification during etching.

Number of e–ALE cycles	Pb$_{UPD}$ charge density (μC/cm^2)
0 (un-etched)	316
2	333
4	351
6	316

4.4 Conclusions

A novel electrochemical atomic layer etching process for Ru is demonstrated. This process consists of three steps: (*1*) surface oxidation of Ru to form a surface monolayer of Ru(OH)$_3$, (*2*) selective etching of this surface monolayer through spontaneous complexation of Ru^{+3} with Cl$^-$, and (*3*) removal of any adsorbed Cl$^-$ for making the Ru surface available for subsequent e-ALE cycles. The three-step process can be repeated cyclically for allowing layer-by-layer etching of Ru. A Ru etch rate of 0.12 nm per e-ALE cycle was determined through performing cross-sectional TEM, suggesting that approximately two e-ALE cycles are required to remove an atomic layer of Ru. Furthermore, electrochemical and microscopic characterizations confirm that surface roughness of Ru is not amplified during etching. Similar to Cu e-ALE, the Ru e-ALE process presented in this chapter utilizes a two-step process where the metal surface monolayer is first oxidized in a surface-limiting manner and is then selectively etched away. This process can be potentially extended to ALE of other metals where true layer-by-layer etching with atomic-scale precision is desired.

5.1 Summary and Conclusions

Novel electrochemical atomic layer etching (e-ALE) processes for Cu and Ru have been developed for atomically precise tailoring of surfaces in advanced nanoelectronics. The e-ALE processes follow the two-step approach of (*1*) surface oxidation followed by (*2*) selective etching of the oxidized surface monolayer. The present study leads to the following key conclusions:

(*i*)   Feasibility of a 'two-cell' Cu e-ALE process was demonstrated. This is the first demonstration of true atomic layer etching of bulk metal films using electrochemical methods. The Cu e–ALE process provides an etch rate of close to one Cu monolayer per e–ALE cycle.

(*ii*) Thermodynamic investigation of surface-limited sulfidization of Cu, a key step in Cu e-ALE, was performed. The reaction at equilibrium follows the Frumkin isotherm. Additionally, an optimal sulfidization potential of –0.74 V is determined. At this potential, a near-complete Cu_2S surface coverage is achieved ensuring that the Cu surface roughness is not significantly amplified during e-ALE.

(*iii*) The two-step e-ALE process was extended to a noble metal – Ru. The Ru e-ALE process can be performed in one electrochemical cell, thereby providing advantages for scale-up. A third step was incorporated to remove any adsorbed Cl⁻ on the Ru surface during the selective etching step, thus ensuring the reproducibility of subsequent e-ALE cycles. The three-step process repeated cyclically allowed for layer-by-layer etching of Ru with an etch rate of 0.12 nm per e-ALE cycle, suggesting that approximately two e-ALE cycles are required to remove an atomic layer of Ru.

5.2 Outlook and Future Work

The e-ALE processes for metals presented in this study can be an attractive alternative to conventional vapor-phase, plasma-assisted etching techniques. However, for application of the e-ALE process to resistive or micro-patterned substrates used in semiconductor manufacturing, having external electrical contact to the substrates can be problematic. This would result in a high ohmic resistance across the high-resistivity metal thin-film, which inhibits the formation of a near-complete oxidized surface monolayer. As shown in the case of Cu e-ALE, surface roughness is amplified during etching if surface coverage of the oxidized monolayer is not close to one. To ensure precise and uniform etching of Cu, an 'electroless' mode of performing the surface oxidation step needs to be developed. Similar to the electroless atomic layer deposition process reported previously,[86] electroless ALE of Cu would require a chemical oxidizing agent that controls the Cu 'mixed' substrate potential in a range where the surface oxidation step is spontaneously facilitated.

Other applications also require atomically precise tailoring of materials, e.g., etching of Si in three dimensional structures.[87] However, drawbacks including surface roughness amplification and aspect ratio dependent etching can result from the continuous etching mechanism used by plasma-assisted techniques. The surface-limiting behavior of surface oxidation reactions used in e-ALE processes can be potentially extended to layer-by-layer etching of other materials including non-metals.

APPENDIX A. Design Guidelines for Tubular Flow-through Electrodes for Use in

Electroanalytical Studies of Redox Reaction Kinetics

Work presented in this appendix is part of an ongoing collaboration between Case Western Reserve University and Lawrence Livermore National Laboratory (LLNL). Experiments were performed by collaborators at LLNL, and modeling was performed at CWRU. Work was performed under contract DE-AC52-07NA27344 and LDRD 20-ERD-056. lLNL-JRNL-819093.

A.1 Introduction

Tubular flow-through electrodes (FTEs), consisting of a single cylindrical pore through which electrolyte is passed, have been extensively used as analytical tools in applications such as electrochemical sensing.[88,89] FTEs are easy to fabricate and use, and they facilitate electrochemical measurements in the presence of convective flow.[90,91] Generally, the direction of convective flow is parallel to the electrical field in FTEs as opposed to the flow-by configuration where flow is perpendicular to the electrical field.[92,93] Due to the high rates of convective mass transfer, FTEs enable detection of electroactive species below 10^{-8} M.[90,94] The growing interest in 'flow-through' and 'flow-by' porous electrodes in modern flow batteries and electrochemical reactors has resurrected the interest in FTEs, which represent an idealized single-pore embodiment of an otherwise complex pore network.[95,96] The transport characteristics of tubular FTEs are well-understood,[91] and they more closely mimic the transport characteristics of a practical porous flow-through electrode than the widely used rotating disk electrode (RDE).[97–99]

The tubular FTE, much like other electrode configurations, can be a reliable analytical tool for precise measurement of electrochemical kinetics only if the electrode provides a sufficiently uniform distribution of the reaction current (rate) over its surface. Current distribution non-uniformity can result from preferential transport (via migration, diffusion or convection) of the reacting species to more accessible portions of the tubular FTE. Alkire and Mirarefi computed the current distribution within FTEs by solving the Laplace's equation and the convection-diffusion equation.[96,100] They showed that, in a long tubular electrode with negligible mass transfer resistance, non-uniform current distribution results when the ohmic resistance is more dominant than the kinetic resistance. This is particularly evident when using resistive electrolytes and studying highly reversible charge-transfer reactions. Ben-Porat et al. extended the analysis to also consider the effect of electrode resistance and showed that uniformity in electroplating can be achieved when the cathodic charge-transfer resistance is more dominant than the metal electrode resistance.[101] Blaedel et al. investigated the use of platinum flow-through electrodes for electroanalytical studies,[90,94,102–104] and recommended the use of FTEs only at low operating currents so as to maintain uniform current distribution. The recent resurgence of tubular FTEs for electroanalytical studies and applications,[105–107] and the general lack of user-friendly guidelines on how to design and operate a tubular FTE for this purpose, motivated us to take a fresh look at this classical electrode geometry and analyze its salient features.

Guidelines for the design and operation of a tubular FTE are developed for its use as an electroanalytical tool for reliable measurement of electrochemical kinetics. This goal is accomplished through a combination of scaling analysis and analytical modeling of the current distribution, which leads to quantitative criterion (based on two dimensionless parameters: Wa_I and Wa_{II}) for selecting the tubular FTE dimensions and operating average current density. The developed guidelines are applied to the design of a graphite tubular FTE and show that they lead to kinetics parameters for the ferri/ferrocyanide redox reaction that are in excellent agreement with those obtained using RDE.

A.2 Experimental Details

A.2.1 Materials

A simple graphite tubular flow-through electrode (FTE) was prepared by drilling a 1-mm-diameter hole ($d = 1$ mm) through a graphite rod (6.15 mm diameter, 99.9995% metals basis, Alfa Aesar). The rod was cut to 4.5 mm in length ($L = 4.5$ mm). A conductive silver wire was glued with conductive epoxy to the outer surface of the graphite electrode in order to make proper electrical contact. Non-conductive epoxy (thickness ≈ 0.02 cm) was applied on all outer surfaces of the electrode. Thus, only the interior surface of the FTE was exposed to the electrolyte. The electrode was glued with non-conductive epoxy to the end of a glass tube (ID ≈ 5.6 mm), which was used in a three-electrode cell described in Fig. A.1. Before electrochemical experiments, the FTE was cured in the oven at 100 °C for > 12 hours and then soaked in isopropyl alcohol (IPA) for 10 minutes. The IPA was allowed to evaporate in the oven for 5 minutes.

The RDE was manufactured from a graphite rod. The graphite rod was shaped with 600-grit silicon carbide paper, polished with 5, 0.3, and 0.05-μm alumina polishing solutions on a micro-fiber polishing cloth, and washed with DI water. The diameter of the RDE was 0.5 cm.

A.2.2 Methods

As shown in Fig. A.1a, the graphite tubular FTE was tested in a custom two-compartment cell with a three-electrode configuration consisting of the FTE as the working electrode, a coiled platinum (Pt) counter electrode, and a saturated Ag/AgCl reference electrode (Pine Research Instrumentation, Inc.). The two compartments were separated by a Nafion membrane to avoid re-oxidation and re-reduction of species electrochemically generated at the Pt counter electrode. The electrolytes consisted of varying equimolar concentrations (1, 0.5, 0.25, and 0.1 mM) of potassium hexacyanoferrate (II) trihydrate $(K_4[Fe(CN)_6] \cdot 3H_2O$, Sigma Aldrich) and potassium hexacyanoferrate (III) $(K_3Fe(CN)_6$, >99.0% purity, Sigma Aldrich) supported by 1 M KCl. The equilibrium potential (E_{eq}) of the working electrode was well-defined, measurable and stable in the presence of equimolar amounts of $[Fe(CN)_6]^{4-}$ and $[Fe(CN)_6]^{3-}$ in the electrolytes. The electrolyte flow direction was upward through the graphite tubular FTE at a flow rate (V_0)

of 5 cm^3/min controlled by a syringe pump (KDS200, KD Scientific). A SP-300 Potentiostat (Bio-Logic Science Instruments, France) was used for electrochemical measurements. Before steady-state polarization experiments were performed on the FTE and RDE, cyclic voltammetry (potential range of −0.4 to 0.2 V vs. Ag/AgCl) at a scan rate of 0.02 V/s was applied to the FTE for 5 cycles in 1 M KCl to reactivate the electrode surface. To obtain steady-state polarization curves, chronoamperometry experiments were performed as follows: the working electrode potential was changed stepwise in increments of 0.01 V and the transient current response at each potential was recorded until steady-state current was achieved. Applied potentials were within the electrochemical stability window of water. Electrochemical impedance spectroscopy (EIS) was performed to determine the ohmic resistance, which was needed for iR_Ω correction. Steady-state polarization measurements and EIS were also performed on the RDE at a rotation rate of 500 rpm. Atomic force microscopy (AFM) analysis was performed to quantitively characterize the surface roughness of the FTE and RDE.

A.3 Results and Discussion

A.3.1 Dimensionless Parameters Representing Transport and Reaction Processes in a Tubular FTE

Scaling analysis is performed for a simplified one-dimensional tubular FTE geometry shown in Fig. A.1 to assess the current distribution uniformity over the FTE surface. For clarity, Table A.1 lists definitions of all the symbols used in this work. For scaling analysis, the assumptions are: (*i*) A high aspect ratio FTE with $L \gg d$; (*ii*) Negligible potential and concentration gradients in the radial direction within the tubular FTE; and (*iii*) Current density on the interior surface of the FTE depends only on the axial position z.

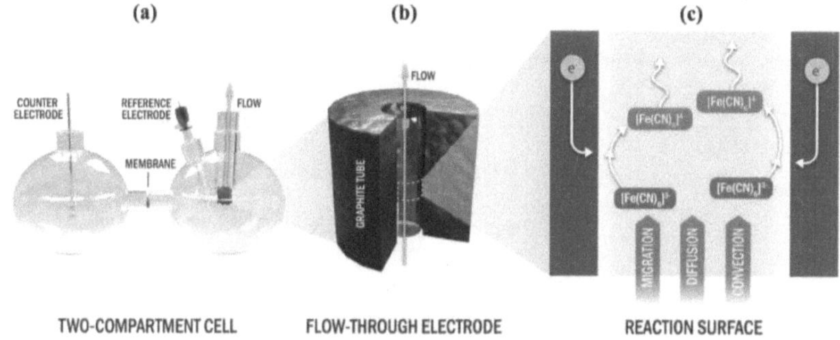

Figure A.1. (a) Schematic representation of a tubular FTE experimental setup. (b) 3-D cross-section showing flow of electrolyte within the FTE in the axial direction. (c) Schematic representation of the various transport and reaction processes within the tubular FTE, assuming a ferri/ferrocyanide redox reaction.

Table A.1. List of symbols and definitions used in developing design guidelines for tubular flow-through electrodes.

A_c	Cross-sectional area of FTE, cm^2
A_s	Interior surface area of FTE, cm^2
C^b	Concentration of equimolar redox species in the bulk electrolyte, M
C_j	Concentration of ferrocyanide (j = II) and ferricyanide (j = III), M
D_j	Diffusion coefficient of ferrocyanide (j = II) and ferricyanide (j = III), cm^2/s
d	Diameter of FTE, cm
E_{eq}	Equilibrium potential of [Fe(CN)₆]$^{4-}$/[Fe(CN)₆]$^{3-}$ redox couple, V

F	Faraday's constant (= 96485 C/mol)
I	Total current at the FTE-electrolyte interface, mA
I_c	Total axial current flowing through the electrolyte within FTE [= $A_c \cdot i_c$], mA
I_s	Total reaction current density on FTE surface [= $A_s \cdot i_s$], mA
i	Current density, mA/cm^2
i_{avg}	Average current density at the FTE-electrolyte interface, mA/cm^2
i_c	Axial current density flowing through the electrolyte within FTE, mA/cm^2
i_k	Kinetic current density, mA/cm^2
i_L	Transport-limited current density, mA/cm^2
i_s	Reaction current density on FTE surface, mA/cm^2
i_0	Exchange current density of the redox reaction, mA/cm^2
L	Length of FTE, cm
m	A parameter characterizing the current distribution [= $\mathrm{Wa_I}^{-0.5}$], dimensionless
n	Number of electrons transferred in the reaction, dimensionless
R	Universal gas constant (= 8.314 J/mol K)
R_a	Activation resistance, Ω
R_{conv}	Convective mass transport resistance, Ω
R_{diff}	Diffusional mass transport resistance, Ω
R_{mt}	Total mass transport resistance, Ω
R_Ω	Ohmic resistance of the electrolyte, Ω
r_{RDE}	Radius of RDE, cm
T	Temperature, K
V_{app}	Applied electrode potential, V

V_0	Flow rate in FTE, cm^3/min
v_0	Average axial flow velocity in FTE [$= 4V_0/\pi d^2$], cm/s
Wa_I	First Wagner number, dimensionless
Wa_{II}	Second Wagner number, dimensionless
z	Axial position coordinate in FTE (inlet: $z = 0$; outlet: $z = L$), cm
α_a	Anodic charge transfer coefficient, dimensionless
η_a	Activation overpotential, V
η_c	Concentration overpotential, V
η_Ω	Ohmic overpotential, V
κ	Electrolyte conductivity, S/cm
μ	A parameter characterizing the current distribution, dimensionless

For scaling analysis, the activation, ohmic and mass transport resistances associated with the tubular FTE were first calculated. The activation resistance (R_a) is the reaction resistance due to irreversible charge transfer at the FTE-electrolyte interface. The ohmic resistance (R_Ω) is the resistance of the electrolyte through which ionic current flows in the presence of an electric field. Finally, the mass transport resistance (R_{mt}) is the resistance associated with diffusional or convective transport of species to the reacting interface.

By definition, the activation resistance (R_a) is the gradient of the activation overpotential vs. current curve calculated at the average value of the current. If we assume Tafel kinetics,[108] then

$$R_\mathrm{a} = \left[\frac{\partial \eta_\mathrm{a}}{\partial I_\mathrm{s}}\right]_{i_\mathrm{s}=i_\mathrm{avg}} = \left(\frac{1}{A_\mathrm{s}}\right)\left[\frac{\partial \eta_\mathrm{a}}{\partial i_\mathrm{s}}\right]_{i_\mathrm{s}=i_\mathrm{avg}} = \left(\frac{1}{\pi d L}\right)\left(\frac{RT}{\alpha_\mathrm{a} F i_\mathrm{avg}}\right) \qquad [\text{A.1}]$$

Similarly, the ohmic resistance (R_Ω) of the tubular FTE can be calculated as:

$$R_\Omega = \frac{\partial \eta_\Omega}{\partial I_\mathrm{c}} = \left(\frac{1}{A_\mathrm{c}}\right)\frac{\partial \eta_\Omega}{\partial i_\mathrm{c}} = \left(\frac{4}{\pi d^2}\right)\left(\frac{L}{\kappa}\right) \qquad [\text{A.2}]$$

In a well-supported electrolyte, the resistance associated with transport of reacting species (R_mt) may be attributed to diffusional or convective transport. Since diffusional or convective transport may occur in tandem in a tubular FTE (resistances in parallel), we can write:

$$\frac{1}{R_\mathrm{mt}} = \frac{1}{R_\mathrm{diff}} + \frac{1}{R_\mathrm{conv}} \qquad [\text{A.3}]$$

The diffusional transport resistance (R_diff) is the gradient of the concentration overpotential vs. current curve:

$$R_\mathrm{diff} = \frac{\partial \eta_\mathrm{c}}{\partial I_\mathrm{c}} = \left(\frac{1}{A_\mathrm{c}}\right)\frac{\partial \eta_\mathrm{c}}{\partial i_\mathrm{c}} = \left(\frac{4}{\pi d^2}\right)\frac{\partial}{\partial i_\mathrm{c}}\left[-\frac{RT}{nF}\ln\left(1 - \frac{i_\mathrm{c} L}{nF D_\mathrm{j} C^b}\right)\right] \qquad [\text{A.4}]$$

For small values of i_c, R_diff can be approximated as:

$$R_{\text{diff}} \approx \left(\frac{4}{\pi d^2}\right)\left(\frac{RTL}{n^2 F^2 D_j C^b}\right) \qquad [\text{A.5}]$$

Similarly, the convective mass transport resistance (R_{conv}) can be determined:

$$R_{\text{conv}} = \left(\frac{1}{A_c}\right)\frac{\partial \eta_c}{\partial i_c} = \left(\frac{4}{\pi d^2}\right)\frac{\partial}{\partial i_c}\left[-\frac{RT}{nF}\ln\left(1 - \frac{i_c}{nF v_0 C^b}\right)\right] \qquad [\text{A.6}]$$

$$R_{\text{conv}} \approx \left(\frac{4}{\pi d^2}\right)\left(\frac{RT}{n^2 F^2 v_0 C^b}\right) \qquad [\text{A.7}]$$

Two dimensionless parameters are defined here: Wa_I, i.e., the classical Wagner number widely used to analyze electrochemical systems,[109–111] and Wa_II, i.e., a modified version of the Wagner number incorporating the effect of mass transport. Wa_I represents the ratio of the activation resistance to the electrolyte ohmic resistance, and Wa_II represents the ratio of the activation resistance to the mass transport resistance:[109,112,113]

$$\text{Wa}_\text{I} = \frac{R_a}{R_\Omega} = \left(\frac{RT\kappa}{\alpha_a F i_{\text{avg}}}\right)\left(\frac{d}{4L^2}\right) \qquad [\text{A.8}]$$

$$\text{Wa}_\text{II} = \frac{R_a}{R_{\text{mt}}} = \left(\frac{n}{\alpha_a}\right)\left[\frac{nF(D_j + L v_0)C^b}{i_{\text{avg}}}\right]\left(\frac{d}{4L^2}\right) \qquad [\text{A.9}]$$

For any practical tubular FTE operating at typical flow rates, $L v_0$ is of the order of 1 cm^2/s and thus $L v_0 \gg D_j$ implying that convection is significantly more effective at transporting reactant species compared to diffusion. Thus, for a typical tubular FTE with convection, Eq. A.9 simplifies to:

$$\mathrm{Wa_{II}} \approx \left(\frac{n}{\alpha_a}\right)\left[\frac{nFLv_0 C^b}{i_{avg}}\right]\left(\frac{d}{4L^2}\right) \qquad\qquad [A.10]$$

Note that $\mathrm{Wa_{II}}$ is inversely correlated to the Thiele modulus (Th) and Damköhler number (Da) commonly found in literature on chemical kinetics.[114,115] From Eqs. A.8 and A.9, it can be concluded that for uniform current distribution in a tubular FTE whereby the overall process is activation controlled (ohmic and transport resistances are small in comparison to the activation resistance), $\mathrm{Wa_I}$ and $\mathrm{Wa_{II}}$ must be large quantities.

Fig. A.2 shows $\mathrm{Wa_I}$ and $\mathrm{Wa_{II}}$ values plotted as a function of the average current density (i_{avg}). For this plot, parameter values, e.g., C^b, n, v_0, α_a, and κ, are reported in Table A.2. These values represent closely the ferri/ferrocyanide reaction on graphite, as discussed below. To achieve large values of $\mathrm{Wa_I}$ and $\mathrm{Wa_{II}}$, i.e., uniform current distribution, Fig. A.2 shows that the tubular FTE must be operated at a low average current density because $\mathrm{Wa_I}$ and $\mathrm{Wa_{II}}$ both depend inversely on i_{avg}. At typical flow rates through the FTE, e.g., $v_0 = 10.6$ cm/s used in Fig. A.2, we observe that $\mathrm{Wa_{II}}$ is very large ($\mathrm{Wa_{II}} > 100$ over the entire range of current densities shown) and that $\mathrm{Wa_{II}} \gg \mathrm{Wa_I}$. This is a manifestation of the fact that convective transport through the tubular FTE considerably lowers its transport resistance compared to its typical ohmic resistance.

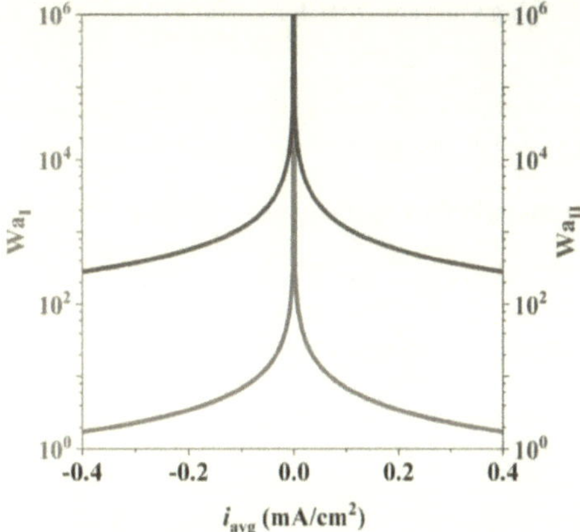

Figure A.2. Wa_I and Wa_{II} calculated for a tubular FTE (using conditions listed in Table A.2) as a function of the average current density. Plot shows that $Wa_{II} \gg Wa_I$ and that large Wa_I (uniform current distribution) requires operation at low average current density. Sign convention used for this figure is that negative currents imply cathodic reaction, and positive currents imply anodic reaction.

Table A.2. Parameters used in scaling analysis and analytical modeling.

Parameter	Value
C^b	1 mM
D_{II}	0.677×10^{-5} cm²/s
D_{III}	0.726×10^{-5} cm²/s
d	0.1 cm
L	0.45 cm
n	1
v_0	10.6 cm/s
α_a	0.5
κ	0.109 S/cm

Parameters represent typical values for ferri/ferrocyanide reaction on a graphite tubular FTE. Diffusion coefficients were taken from Ref. 116 and electrolyte conductivity was taken from Ref. 117.

A.3.2 Quantifying Current Distribution Non-Uniformity via Analytical Modeling

In this section, analytical modeling is performed to quantify the extent of current distribution non-uniformity as a function of Wa_I and Wa_{II}. The current distribution over the tubular FTE surface was modeled for the simplified axisymmetric geometry shown in Fig. A3a. An analytical expression for the secondary current distribution (small Wa_I, large Wa_{II}) for this geometry, assuming linearization of the kinetics near i_{avg},[109] was provided by Akolkar:[118]

$$\frac{i_s}{i_{avg}} = m \left[\frac{e^{m\left(1-\frac{z}{L}\right)} + e^{-m\left(1-\frac{z}{L}\right)}}{e^m - e^{-m}} \right] \qquad [A.11]$$

Here, $m = Wa_I^{-0.5}$. When ohmic resistance is negligibly small (large Wa_I) but diffusion resistance is large ($v_0 = 0$ providing small Wa_{II}), the current distribution obeys:[119]

$$\frac{i_s}{i_{avg}} = \mu \left[\frac{e^{\mu\left(1-\frac{z}{L}\right)} + e^{-\mu\left(1-\frac{z}{L}\right)}}{e^\mu - e^{-\mu}} \right] \qquad [A.12]$$

where μ is related to Wa_{II} as: $\mu \cdot \tanh(\mu) = (\alpha_a \cdot Wa_{II})^{-1}$ for a one-electron transfer reaction (n = 1). The analytically modeled current profiles over the FTE surface at various Wa_I and Wa_{II} values are shown in Fig. A.3b. In the absence of flow ($v_0 = 0$ cm/s) and at very low average current density, Wa_I is large (negligible ohmic resistance compared to activation resistance) but Wa_{II} is small (substantial diffusional mass transport resistance compared to activation resistance). This leads to significant non-uniformity in the current distribution (Fig. A.3b) with less current able to penetrate deep into the FTE due to diffusion limitations in the absence of flow. The presence of flow through an FTE is thus of paramount importance in eliminating diffusion limitations and achieving large Wa_{II} as seen in Eq. A.9. In the presence of flow through the tubular FTE ($v_0 = 10.6$ cm/s), Wa_I and Wa_{II} values decrease as i_{avg} increases. Increase in i_{avg} makes the current increasingly non-uniform (Fig. A.3b). In all cases, $Wa_{II} \gg Wa_I$ due to the fact that mass transport resistance is nearly eliminated by the flow attributes of the FTE. Notably, the ratio i_s/i_{avg} (which is maximum at the FTE entrance $z/L = 0$) is equal to or less than 1.05 for $Wa_I > 6$. We take $i_s/i_{avg} \leq 1.05$ to represent relatively uniform current distribution and thus the criterion $Wa_{II} \gg Wa_I > 6$ becomes the key design criterion to ensure reasonably uniform current distribution across the tubular FTE.

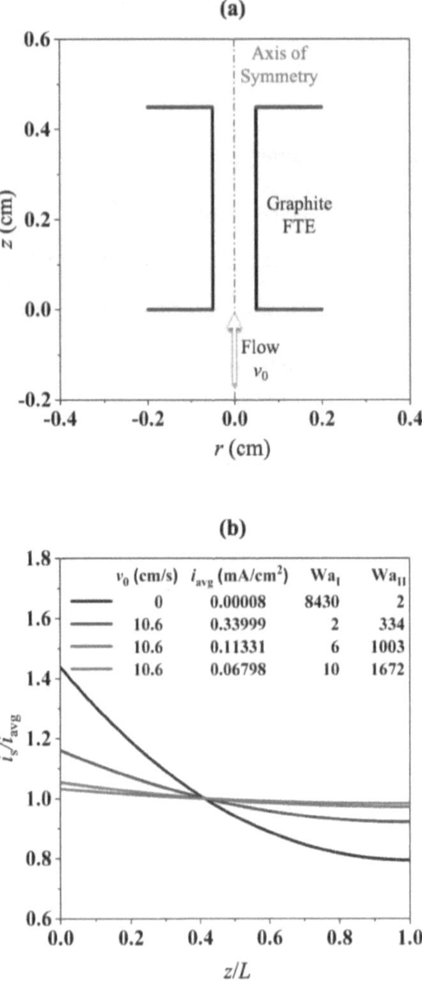

Figure A.3. (a) Schematic representation of the 1-D FTE geometry used for analytical modeling of the current distribution (Eqs. A.11 and A.12 in the text). (b) Current distribution across the tubular FTE surface at various v_0 and i_{avg}. Current distribution is uniform ($i_s/i_{avg} \leq 1.05$) at $Wa_{II} \gg Wa_I > 6$.

A.3.3 FTE Operating Conditions that Yield Reliable Kinetics Data

The above evaluation of current distribution using scaling analysis and analytical modeling provided key design criterion for tubular FTEs: $Wa_{II} \gg Wa_I > 6$. Plots of Wa_I and Wa_{II} at various i_{avg}, d, and L values are shown in Fig. A.4. Analysis presented in Fig. A.4 assumes a constant flow velocity ($v_0 = 10.6$ cm/s). To achieve large values of Wa_I and Wa_{II}, FTE must be operated at a small average current density (i_{avg}) and must have a large diameter (d) or a small length (L). Fig. A.4 shows that, to achieve uniform current distribution, i.e., $Wa_{II} \gg Wa_I > 6$, the FTE must be designed to have a diameter $d \geq 0.2$ cm and be operated at an average current density $i_{avg} \leq 0.23$ mA/cm^2 when L is fixed at 0.45 cm. Alternatively, the tubular FTE must have a length $L \leq 1$ cm and be operated at $i_{avg} \leq 0.02$ mA/cm^2 when d is fixed at 0.1 cm. In Fig. A.4, conditions below the horizontal solid line ($Wa_{II} = Wa_I = 6$) should be avoided as these would lead to non-uniform current distribution.

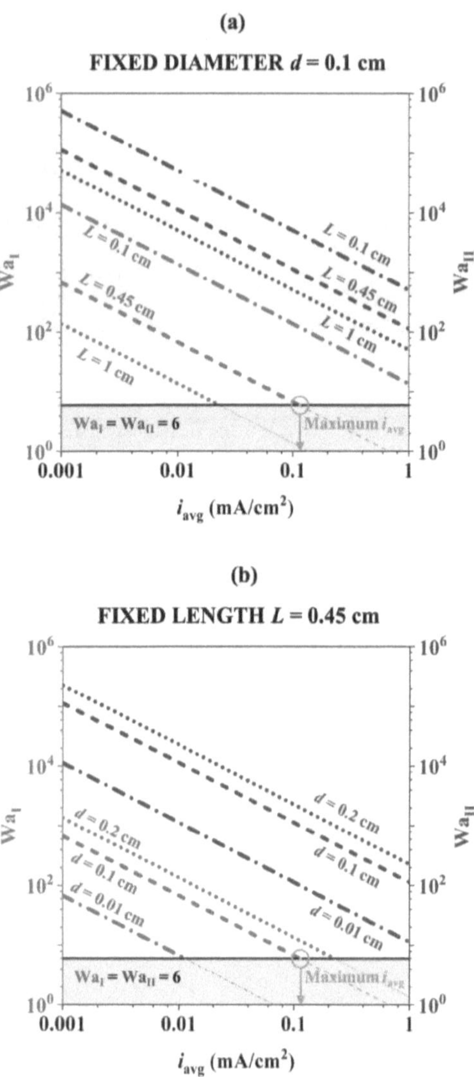

Figure A.4. Effect of FTE parameters (i_{avg}, d, and L) on Wa_I and Wa_{II}: (a) Constant L, varying d and (b) Constant d, varying L. Thicker lines represent design regime where $Wa_{II} \gg Wa_I > 6$.

97

Guided by Fig. A.4, we now investigate the ferri/ferrocyanide reaction kinetics on a graphite tubular FTE. A tubular FTE setup is constructed as described under 'methods' with $L = 0.45$ cm and $d = 0.1$ cm. Fig. A.5a shows steady-state polarization curves measured on a graphite FTE at various equimolar concentrations of $K_3Fe(CN)_6$ and $K_2Fe(CN)_6$. A stable equilibrium potential (E_{eq}) was established at each $[Fe(CN)_6]^{4-}$/$[Fe(CN)_6]^{3-}$ concentration (e.g., $E_{eq} = 0.33$ V vs. Ag/AgCl at $C_{II} = C_{III} = 1$ mM). At large overpotentials ($V_{app} - IR_\Omega - E_{eq} > 0.2$ V), mass transport limitations are observed in Fig. A.5a. Below the mass transport limit, the anodic voltammogram in the Tafel range can be described by:

$$i_{avg} = i_0 \left(1 - \frac{i_{avg}}{i_L}\right) \exp\left[\frac{\alpha_a F}{RT}\left(V_{app} - IR_\Omega - E_{eq}\right)\right] \qquad [A.13]$$

Tafel plots (Fig. A.5b) were constructed where the kinetic current density $i_k = |i_{avg}\,[i_L/(i_L - i_{avg})]|$ was plotted vs. $V_{app} - IR_\Omega - E_{eq}$. For Tafel kinetics, semi-log plots of i_k vs. $V_{app} - IR_\Omega - E_{eq}$ show a region with linearity because:

$$\log_{10} i_k = \log_{10} i_0 + \left[\frac{\alpha_a F}{(2.303)RT}\right]\left(V_{app} - IR_\Omega - E_{eq}\right) \qquad [A.14]$$

In Fig. A.5b, the data points chosen from the raw polarization plot (Fig. A.5a) meet the design criterion: $Wa_{II} \gg Wa_I > 6$. This ensures that non-uniform current distribution does not convolute Tafel analysis. As seen in Eq. A.14 and its application to Fig. A.5b, the Tafel slope provided the anodic charge transfer coefficient: $\alpha_a = 0.58$, and the Y-intercept at

$V_{app} = E_{eq}$ provided the exchange current density: $i_0 = 0.046$ mA/cm^2 (at equimolar concentration of 0.25 mM). Similar analyses were performed on a graphite RDE: the raw polarization data (Fig. A.5c) was screened for uniform current distribution, followed by Tafel analysis using Eq. A.14. Analysis yielded $\alpha_a = 0.56$ and $i_0 = 0.052$ mA/cm^2 for equimolar concentration of 0.25 mM. Good agreement between the two methods is observed in Table A.3 which compares the kinetics parameters α_a and i_0 determined using FTE and RDE over a range of concentrations. Note that, for the RDE (Fig. A.5c), the conventional Wagner number takes the form:[112]

$$Wa_{I\_RDE} = \left(\frac{RT\kappa}{\alpha_a F i_{avg}} \right) \left(\frac{4}{\pi r_{RDE}} \right) \qquad [A.15]$$

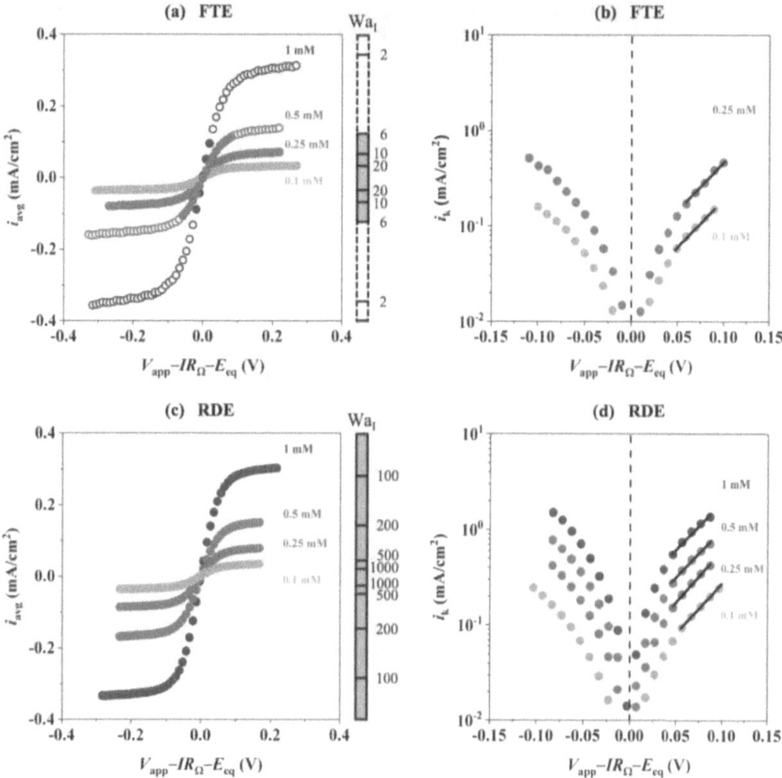

Figure A.5. (a) Steady-state polarization for the $[Fe(CN)_6]^{4-} \rightleftharpoons [Fe(CN)_6]^{3-}$ redox reaction on a graphite tubular FTE at various equimolar concentrations of $[Fe(CN)_6]^{4-}$ and $[Fe(CN)_6]^{3-}$. Filled data points meet the criterion $Wa_{II} \gg Wa_I > 6$ and were selected for Tafel plots. (b) Tafel analysis of data selected from (a) leading to electrochemical kinetics parameters. The Tafel regime was selected such that $i_0 < i_{avg} < 0.9i_L$. (c) and (d) represent similar analysis as (a) and (b) but using RDE at 500 rpm. Table A.3 summarizes kinetics parameters i_0 and α_a determined using FTE and RDE.

Table A.3. Kinetics parameters (α_a, i_0) at various concentrations (C^b) of $[Fe(CN)_6]^{4-}$ and $[Fe(CN)_6]^{3-}$ measured using FTE and RDE.

C^b (mM)	FTE		RDE	
	α_a	i_0 (mA/cm^2)	α_a	i_0 (mA/cm^2)
1	–	0.195	0.50	0.208
0.5	–	0.082	0.52	0.100
0.25	0.58	0.046	0.56	0.052
0.1	0.55	0.021	0.60	0.025

α_a was not available at $C^b = 1$ and 0.5 mM for the FTE due to unavailability of data points that satisfy $Wa_I > 6$ and lie within the Tafel regime. Therefore, i_0 at $C^b = 1$ and 0.5 mM was determined in the 'linear' regime ($V_{app} - IR_\Omega - E_{eq} < 0.04$ V) where $i_{avg} = i_0 \cdot nF(V_{app} - IR_\Omega - E_{eq})/RT$.

From the dependence of the exchange current density on C^b (Table 3), the standard rate constant can be determined:[108]

$$i_0 = nFk_0C^b \qquad [A.16]$$

Fig. A.6 shows a plot of i_0 vs. C^b and contains kinetics data acquired on the tubular FTE as well as the RDE following the design criterion ($Wa_I > 6$). From the slopes in Fig. A.6, the standard rate constants (k_0) were extracted: for FTE, $k_0 = 1.95 \times 10^{-3}$ cm/s, and for RDE, $k_0 = 2.14 \times 10^{-3}$ cm/s. While these kinetics constants are already in very good agreement (within 10%), the slightly higher apparent rate constant measured on the RDE may be attributed to its elevated roughness and thus higher electroactive surface area confirmed by atomic force microscopy (AFM) as shown in Fig. A.7. AFM measurements indicate that the surface area is 6% higher than the projected area for FTE and 22% for RDE. The 15%

difference between the true surface areas of RDE and FTE partially accounts for the

difference between the k_0 values determined using RDE and FTE. The rate constant value

is in the same general range as values reported previously.[120,121] The differing

electrocatalytic properties of graphite electrodes due to oxygenation[122] and edge effects[123]

result in a wide range of rate constant values. Therefore, consistent preparation and

treatment of graphite electrodes is crucial in obtaining reproducible kinetics parameters.

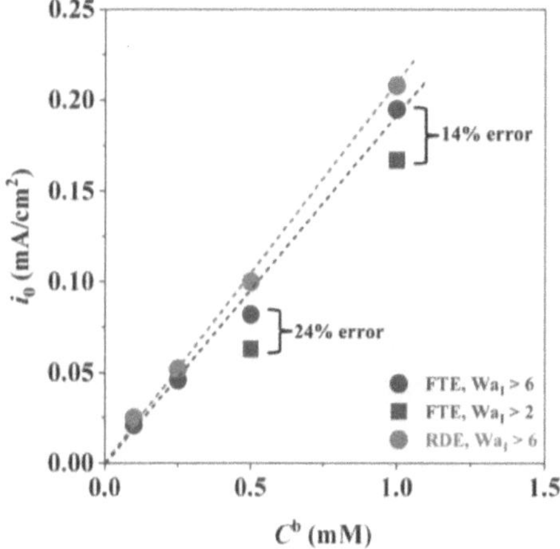

Figure A.6. Exchange current density (i_0) as a function of equimolar concentrations of [Fe(CN)$_6$]$^{4-}$ and [Fe(CN)$_6$]$^{3-}$ for graphite tubular FTE (*blue*) and RDE (*red*). While FTE and RDE based kinetics are in reasonable agreement when using Wa$_I$ > 6 as the FTE design criterion, significant error (14–24%) is introduced in i_0 when applying a weaker design criterion of Wa$_I$ > 2.

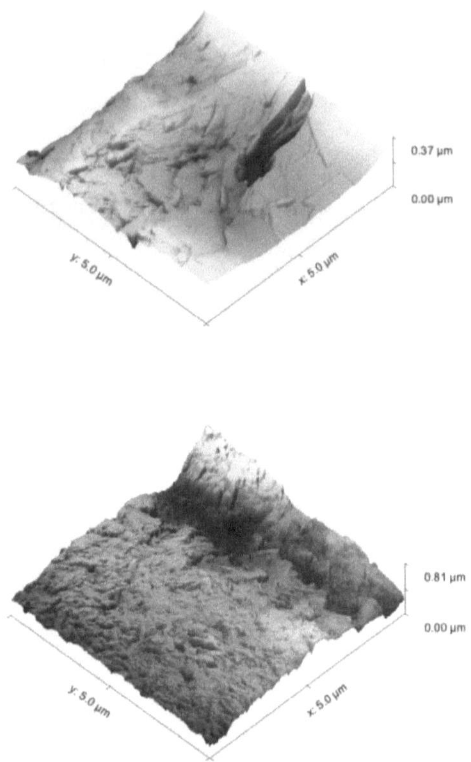

Figure A.7. 3-D AFM images of graphite tubular FTE (*top*) and graphite RDE (*bottom*). The relationships between surface area (SA) and projected area (PA) for the FTE and RDE are as follows: FTE: SA = 1.06 × PA; RDE: SA = 1.22 × PA.

To demonstrate the importance of strict adherence to the design criterion ($Wa_I > 6$) when conducting kinetics analysis, we evaluated the error in i_0 introduced when deviating from the design criterion. In Fig. A.5, if $Wa_I > 2$ is followed as the selection criterion for Tafel analysis, we get $i_0 = 0.167$ mA/cm^2 at $C^b = 1$ mM and 0.067 mA/cm^2 at $C^b = 0.5$ mM.

As seen in Fig. A.6, these values are 14–24% lower than those for the correct criterion ($Wa_I > 6$, which ensures highly uniform current distribution). Analogous to well-established analysis of the disk electrode,[124] a mathematical treatment of the magnitude of error in kinetics measurement using FTE and its dependence on the FTE current distribution non-uniformity will be reported in a subsequent contribution.

A.4 Conclusions

Quantitative guidelines for the design and operation of a tubular FTE are developed. It is demonstrated that, an FTE with $Wa_{II} \gg Wa_I > 6$ provides a uniform current distribution. This condition can be satisfied by choosing appropriate FTE dimensions (large d, small L), and by operating the FTE at low average current densities and high flow rates. The process for choosing the correct parameters for the FTE is summarized in Figure A.8. Such optimized FTE design and operation guarantees that its use as an electroanalytical tool leads to reliable electrochemical kinetics data. This is demonstrated for the $[Fe(CN)_6]^{4-} \rightleftharpoons [Fe(CN)_6]^{3-}$ reaction over a graphite FTE, and results are compared rigorously to those obtained on a conventional RDE. Operation away from the design criterion ($Wa_{II} \gg Wa_I > 6$) are shown to introduce significant errors.

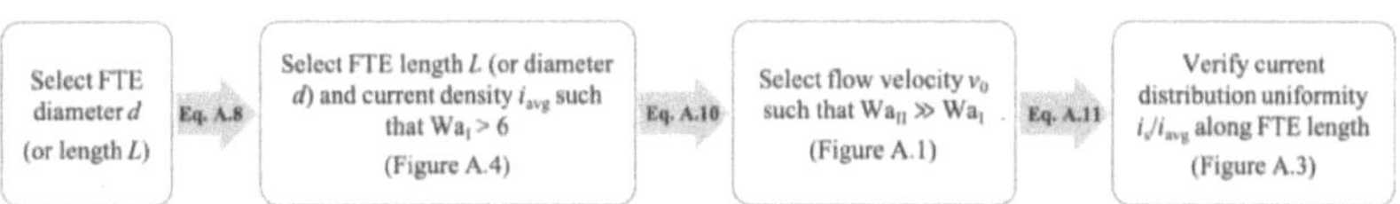

Figure A.8. Process flow chart for the design and operation of a tubular flow-through electrode for electrochemical kinetics studies.

(114) Fogler, H. S. *Elements of Chemical Reaction Engineering*, 4th Edition.; Prentice. Hall: Upper Saddle River, NJ, 2005.

(115) Hill, C. G.; Root, T. W. *Introduction to Chemical Engineering Kinetics and Reactor Design*, 2nd edition.; Wiley, 2014.

(116) Konopka, S. J.; McDuffie, Bruce. Diffusion Coefficients of Ferri- and Ferrocyanide Ions in Aqueous Media, Using Twin-Electrode Thin-Layer Electrochemistry. *Anal. Chem.* **1970**, *42*, 1741–1746.

(117) *CRC Handbook of Chemistry and Physics, 99th Edition*; Rumble, J. R., Ed.; CRC Press, 2018.

(118) Akolkar, R. Analysis Of The 'Bottom–Up' Fill During Copper Metallization Of Semiconductor Interconnects, Ph.D. book, Case Western Reserve University, 2005.

(119) Akolkar, R. Characterizing Transport Limitations during Copper Electrodeposition of High Aspect Ratio Through-Silicon Vias. *ECS Electrochem. Lett.* **2012**, *2*, D5.

(120) J. Slate, A.; C. Brownson, D. A.; Dena, A. S. A.; C. Smith, G.; A. Whitehead, K.; E. Banks, C. Exploring the Electrochemical Performance of Graphite and Graphene Paste Electrodes Composed of Varying Lateral Flake Sizes. *Physical Chemistry Chemical Physics* **2018**, *20*, 20010–20022.

(121) Nwamba, O. C.; Echeverria, E.; Yu, Q.; Raja, K. S.; McIlroy, D. N.; Shreeve, J. M.; Aston, D. E. Increased Electron Transfer Kinetics and Thermally Treated Graphite Stability through Improved Tunneling Paths. *J Mater Sci* **2020**, *55*, 11411–11430.

(122) Ji, X.; Banks, C. E.; Crossley, A.; Compton, R. G. Oxygenated Edge Plane Sites
Slow the Electron Transfer of the Ferro-/Ferricyanide Redox Couple at Graphite
Electrodes. *ChemPhysChem* **2006**, *7*, 1337–1344.

(123) Banks, C. E.; Davies, T. J.; Wildgoose, G. G.; Compton, R. G. Electrocatalysis at
Graphite and Carbon Nanotube Modified Electrodes: Edge-Plane Sites and Tube
Ends Are the Reactive Sites. *Chem. Commun.* **2005**, No. 7, 829–841.

(124) West, A. C.; Newman, J. Corrections to Kinetic Measurements Taken on a Disk
Electrode. *J. Electrochem. Soc.* **1989**, *136*, 139.

9 783384 255990